Plato's Press

CHICAGO, ILLINOIS

THE LIBRARY
OF Congress HAS
CATALOGUED THIS TEXT AS FOLLOWS
Registration: XXX

THE ART AND SCIENCE OF
BREWING
Traditional Techniques, Modern Innovations, and Future Trends

Description: Electronic File (eService)
Copyright: Jeffrey Wallace Bulger
Last Revision: 4/12/2025
Authorship: Jeffrey Wallace Bulger
Rights and Permissions: Jeffrey Wallace Bulger

ISBN 978-1-62252-028-2

9 781622 520282

Book Description

THE ART AND SCIENCE OF BREWING

Traditional Techniques, Modern Innovations, and Future Trends

Jeffrey W. Bulger

In the dance of yeast and malt, tradition whispers while science sings.

Unlock the secrets of brewing mastery with this comprehensive guide that bridges the gap between time-honored traditions and cutting-edge advancements. "**The Art and Science of Brewing**" invites you on a captivating journey through the rich history, intricate science, and innovative future of beer crafting.

Discover:

- **Historical Insights:** Delve into the evolution of brewing from ancient origins to the modern era, understanding how historical practices have shaped today's techniques.
- **Scientific Foundations:** Explore the chemistry and biology that transform simple ingredients into complex flavors, gaining a deeper appreciation for each step of the brewing process.
- **Practical Techniques:** Master advanced methods in mashing, fermentation, barrel aging, sour beer production, and more, with detailed guidance suited for both novices and seasoned brewers.
- **Innovative Trends:** Stay ahead of the curve by learning about emerging styles, sustainability practices, and technological innovations that are redefining the brewing industry.

- **Problem-Solving Strategies:** Enhance your critical thinking skills using rational thought, empirical observation, and the scientific method to troubleshoot and refine your brewing practices.

About the Author

Dr. Jeffrey W. Bulger brings decades of experience and a profound passion for brewing to this definitive guide. With a Ph.D. in Philosophy specializing in Biomedical Ethics from the University of Tennessee and a Bachelor's degree in Petroleum Engineering/Geology from the University of North Dakota, Dr. Bulger uniquely bridges the worlds of science, ethics, and craftsmanship. A dedicated homebrewer and certified beer judge through the Beer Judge Certification Program (BJCP), he combines academic rigor with practical expertise. Known for his commitment to mentoring and education, Dr. Bulger has guided numerous students and contributed significantly to the fields of clinical ethics and research. His scholarly work and hands-on experience make this book an indispensable resource for brewers of all levels.

Embark on a journey through the art and science of brewing. Whether you're crafting your first batch in your kitchen or managing a commercial brewery, this book is your ultimate guide to mastering the craft of exceptional beer.

> *Brewing is where matter meets meaning — where ingredients transform not just through heat and time, but through the brewer's intention, inquiry, and care.*

***

Table of Contents

Introduction: Brewing—Art, Science, & Innovation

Brewing bridges the ancient and the modern, crafting a legacy of tradition and innovation in every glass.

Imagine standing in a bustling brewery, the warm, inviting aroma of malt and hops filling the air, as the soft hum of modern brewing equipment creates a symphony of progress and tradition. The brewer stands at the helm, blending ancient wisdom with modern technology, guided by the principles that have shaped the craft for millennia. The brew kettle hums, the fermenters bubble, and the aroma of hops fills the air. Each batch of beer is a testament to the brewer's skill, creativity, and dedication, a reflection of the journey that began thousands of years ago and continues to evolve. The brewer carefully monitors the process, making adjustments as needed, always striving for perfection. They draw on their knowledge of chemistry and microbiology, but also rely on their senses and intuition. The art of brewing is in the balance between precision and creativity, between science and tradition.

As we embark on this exploration, we celebrate the rich tapestry of history, the vibrant creativity of the present, and the boundless potential of the future. Each chapter in this book is designed to deepen your understanding and enhance your brewing skills. From the fundamental ingredients of malt, hops, yeast, and water, to the intricate processes of mashing, lautering, boiling, fermentation, and packaging, we will cover the essential techniques and advanced methods that define the craft. We will delve into specialized topics such as barrel aging, sour beer production, and sustainability practices, offering insights and practical tips to elevate your brewing.

In the following chapters, you will find detailed explanations of key concepts, step-by-step instructions, and practical advice drawn from both traditional methods and modern innovations. We will explore the science behind brewing, demystifying the chemical and biological processes that transform simple ingredients into complex and flavorful beers. By understanding the principles of rationalism, empiricism, and the scientific method, you will be equipped to approach brewing with a critical eye and a creative spirit, continually refining your techniques and pushing the boundaries of what is possible.

For beer enthusiasts, this book provides a deeper appreciation of the artistry and science behind each brew, enriching your tasting experiences and broadening your knowledge of beer's rich history and diverse styles. For homebrewers, this book provides a wealth of information to help you improve your craft, from selecting the best ingredients to mastering advanced brewing techniques. Whether you are brewing your first batch or experimenting with new styles, you will find guidance and inspiration to achieve consistently high-quality results.

For professional brewers, this book offers insights into the latest trends and technologies, helping you stay at the forefront of the industry. By embracing innovation and sustainability, you can enhance your brewery's efficiency and product quality while contributing to a more environmentally responsible industry.

Brewing is a journey of continuous discovery and innovation. Each batch you brew stands as a testament to your creativity and dedication. This book aims to inspire you to explore new techniques, embrace the science behind the craft, and ultimately create exceptional beers that bring joy to those who taste them. Raise your glass to the past, present, and future of brewing. May your brewing adventures be filled with learning, creativity, and the satisfaction of crafting something truly special.

Cheers!

To brew is to practice an ethics of creation—where care, patience, and precision are not just means to an end, but moral acts in themselves. In shaping the brew, the brewer shapes the world—reminding us that even in small crafts, how we do something matters as much as what we make.

***

Chapter 1: The Alchemy of Brewing Through Time

Brewing history is the alchemy of time, where ancient traditions and scientific innovations ferment the future.

Brewing is a form of modern alchemy, a transformative art where simple ingredients—water, malt, hops, and yeast—are turned into the golden elixir we call beer. Just as ancient alchemists sought to transmute base metals into gold, brewers blend precise chemical processes with creative techniques to craft unique and flavorful golden brews. This journey is not merely about combining ingredients; it's about harnessing centuries of accumulated knowledge, embracing both tradition and innovation, and applying critical thinking to push the boundaries of what is possible.

Imagine stepping into an ancient Sumerian village around 5,000 BCE, where the air is filled with the earthy aroma of barley being brewed into a rich, nourishing beverage. Women, known as "sabitum," are the primary brewers, transforming grains into a staple of daily life. Their expertise in fermentation and recipe development is essential to the well-being and culture of their society. The "Hymn to Ninkasi," dating back to around 1800 BCE, serves both as a prayer to the goddess of beer and as a detailed brewing recipe, highlighting the empirical knowledge and systematic approach these early brewers employed.

From the fertile lands of Mesopotamia to the banks of the Nile in Egypt, beer has been more than just a drink; it has been a symbol of community, a catalyst for social and economic development, and a reflection of human ingenuity. In ancient Egypt, women were also central to brewing, with goddesses like Hathor embodying the cultural significance of beer. Brewing was intertwined with daily life, religious rituals, and even medicine.

The story of beer is as old as civilization itself—a tale interwoven with the rise and fall of empires, the growth of societies, and the evolution of science and technology. Throughout this journey, the principles of rationalism and empiricism have played pivotal roles. Ancient brewers relied on careful observation and experimentation—inductive reasoning—to refine their techniques. Over time, brewing practices were formalized through deductive reasoning, applying general principles to achieve consistent results. The scientific method emerged as a cornerstone of brewing advancements, where observation, hypothesis formation, experimentation, analysis, replicability, and peer review drive innovation.

As brewing spread across continents and ages, it evolved through the contributions of diverse cultures and individuals. In medieval Europe, brewing became a profession, with both men and women participating, although women were gradually marginalized due to social and economic shifts. The Reinheitsgebot (ryne-HYTS-guh-boht) in 1516 introduced regulations that reflected rationalist principles, standardizing ingredients to ensure quality.

In the modern era, we continue to build upon this rich heritage. The craft beer movement has witnessed a resurgence of creativity and diversity, with brewers pushing the boundaries of flavor and technique. Women are reclaiming their historical place in brewing, enriching the industry with fresh perspectives and innovations. Technological advancements allow for precise control over the brewing process, yet the essence remains the same: transforming simple ingredients into something extraordinary.

As we embark on this exploration of brewing's past and present, this book aims to provide you with the foundational knowledge and structured approaches necessary to become a critical-thinking brewmaster. By intertwining historical developments with philosophical and methodological frameworks, we not only trace how rationalism, empiricism, and the scientific method have advanced brewing but also engage you in developing the cognitive critical thinking skills essential for innovation and problem-solving.

Brewing is a dynamic and evolving art, ever adapting to new challenges and opportunities. The alchemy of brewing is not confined to the past; it's a living practice that thrives on curiosity, experimentation, and the courage to question and refine existing methods. By fostering a spirit of exploration and a commitment to continual improvement, you too can participate in this modern alchemy, transforming simple ingredients into golden brews that honor the past while embracing the future.

This journey through time is more than a historical account; it's an invitation to engage with the principles that have shaped brewing and to apply them in your own practice. Whether you're an aspiring brewer, a seasoned professional, or a beer enthusiast, the fusion of historical wisdom and modern ingenuity awaits you. Together, we'll delve into the art and science of brewing, exploring how critical thinking, creativity, and a deep understanding of both tradition and innovation can lead to extraordinary creations in the world of beer.

Ancient Origins and the Role of Women

Our journey begins in the fertile lands of Mesopotamia, the ancient region between the Tigris and Euphrates rivers, where the earliest known brewing practices emerged around 5,000 BCE. Among the Sumerians, beer was not merely a beverage but a central part of daily life and culture. Brewing was deeply embedded in their society, with women playing a pivotal role in its development.

In Sumerian society, women were the primary brewers, known as "sabitum." They were responsible for producing beer for both family and community consumption. Beer was a staple in the Sumerian diet, consumed by people of all ages and social classes due to its nutritional value and safety compared to potentially contaminated water sources. The prominence of women in brewing is evidenced by historical documents, such as clay tablets containing recipes and hymns.

One of the most significant pieces of Sumerian literature is the "Hymn to Ninkasi," dating back to around 1800 BCE. Ninkasi was the Sumerian goddess of beer, and this hymn serves both as a praise to her and as a detailed brewing recipe. The hymn outlines the brewing process step by step, highlighting the empirical knowledge the Sumerians had developed through careful observation and experimentation with ingredients and techniques. It also underscores the role of women as both practitioners and custodians of this essential craft. Ninkasi was revered not only as a deity but also as a symbol of women's skill and importance in society.

Similarly, in the ancient civilization of Egypt, beer was not just a beverage but a vital part of the diet, enjoyed by people of all social classes. Egyptian brewing dates back to at least 3,500 BCE, as evidenced by archaeological findings of brewing facilities and residues in pottery jars. Women were actively involved in brewing, which was integral to daily life and religious rituals.

Egyptian brewers used barley, emmer wheat, and dates to create a variety of beer styles. Emmer wheat, also known as farro, is an ancient grain that was one of the first domesticated crops in the Near East. It has a rich, nutty flavor profile and, in brewing, enhances the complexity of the malt profile, contributing earthy, slightly sweet, and nutty notes. Proper milling and mashing techniques were essential to fully extract its flavors and fermentable sugars.

The Egyptian goddess Hathor was associated with beer, fertility, music, and joy, highlighting the cultural significance of women in brewing. One famous myth involves Hathor, known as the "Lady of Drunkenness," who was appeased with red beer dyed to resemble blood, saving humanity from her wrath. This story emphasizes the divine connection between beer, women, and the sustenance of life.

Brewing in ancient Egypt was sometimes conducted on a large scale, especially for temple offerings and during festivals. Women, both in households and as part of temple staff, were key participants in these brewing operations. Some women held titles such as “She Who Oversees the Beer” and “Mistress of the House,” indicating positions of authority and specialized knowledge in brewing techniques.

Before the widespread use of hops, ancient brewers in various regions employed herbs and spices to flavor and preserve their beer. In Mesopotamia and Egypt, ingredients like dates, honey, spices, and herbs were added to enhance flavor and preservation. In later periods and different regions across Europe, a mixture known as gruit was commonly used. Gruit was a blend of herbs such as sweet gale (bog myrtle), yarrow, wild rosemary, juniper berries, and other botanicals. Each brewer or region had its proprietary recipe, resulting in a wide variety of flavors and aromas. The use of gruit was not only for flavor but also for its preservative qualities and sometimes for medicinal or mystical properties attributed to the herbs.

Beer was consumed during meals, used in religious rituals, and even offered as wages to workers, including those who built the pyramids. The Ebers Papyrus, dating to around 1550 BCE, contains medical recipes that include beer as an ingredient, indicating its medicinal uses. Egyptian tomb paintings and inscriptions provide detailed depictions of the brewing process, revealing a systematic approach that combined rational planning with empirical methods to achieve consistent results.

The involvement of women in brewing in these ancient societies reflects a recognition of their skills and contributions to essential aspects of life. Brewing was not merely a domestic chore but a respected craft requiring knowledge of ingredients, fermentation processes, and flavoring methods. Women were innovators, experimenting with different grains, fruits, and herbs to create various beer styles suited for everyday consumption and ceremonial purposes.

Their central role in brewing also had economic implications. In both Sumeria and Egypt, surplus beer was sold or traded, and women often managed these transactions, contributing to the household economy. This economic participation afforded women a degree of autonomy and influence within their communities.

Recognizing the contributions of women in ancient brewing provides a fuller understanding of the origins of brewing and the integral part they played in its development. It highlights how brewing was not only a craft but also a means of empowerment and societal contribution for women in ancient civilizations.

Medieval Europe

As Europe entered the medieval period, monasteries emerged as centers of brewing innovation and excellence. Monastic breweries, operated by monks in medieval Europe, were known for their systematic and innovative approaches to brewing, contributing to the advancement of brewing techniques. Monastic communities, following the Rule of St. Benedict from the 6th century, brewed beer to ensure the monks' sustenance and health. The Rule emphasizes a balanced life of prayer, work, and community. Monks brewed beer to support their communities and welcome guests, adhering to principles of diligence and hospitality. This tradition of quality brewing continues in many monastic breweries today.

During this time, the use of hops began to spread across Europe. The earliest documented mention of hops in brewing comes from 822 AD, when a Carolingian Abbot named Adalhard of the Benedictine monastery of Corbie in France mentioned hop gardens in his writings. He noted the cultivation of hops in monastic gardens, although it is unclear whether they were used for brewing or medicinal purposes.

A significant early reference to hops in brewing is found in the works of Hildegard of Bingen (1098–1179), a German Benedictine abbess, writer, composer, and herbalist. Around 1150 AD, in her book "*Physica*" (also known as *Liber Simplicis Medicinae*), she wrote about the preservative qualities of hops in beer:

> *Hops, when put in beer, stops putrefaction and lends longer durability.*

This indicates an awareness of hops' ability to prolong the shelf life of beer, highlighting their antibacterial properties.

By the 13th century, the use of hops in brewing had become more widespread in regions like the Hanseatic League cities of Northern Germany and the Low Countries (modern-day Netherlands and Belgium). Hopped beer allowed for longer storage and transport, which was advantageous for trade. Cities like Hamburg became renowned for their hopped beers, exporting them across Europe.

However, the transition from gruit to hops was not merely a matter of improving beer quality; it was deeply entwined with political and economic factors. The production and sale of gruit were often controlled by local authorities or the Church, who held the gruitrecht (gruit right). This monopoly allowed them to tax brewers on gruit sales, making it a significant source of revenue. The gruitrecht granted exclusive rights to produce and sell gruit, and brewers were required to purchase it from the holders of this right.

As hops became more popular, they threatened the established economic interests tied to gruit. Hops could be cultivated more widely and did not require purchasing from entities controlling gruit. This allowed brewers to bypass taxes associated with gruitrecht, reducing costs and challenging the monopolies held by the Church and local authorities. In response, those profiting from gruit sales opposed the shift to hops, sometimes influencing local laws to restrict their use.

In England, hops were introduced in the late 14th or early 15th century. Initially, there was significant resistance to hopped beer. Traditional English ale was unhopped and flavored with gruit. Hopped beer, was referred to as "beer" to distinguish it from traditional "ale" which was flavored with gruit. Hopped "beer" was viewed as a foreign innovation and was sometimes met with legal opposition. In 1519, for example, the city of London banned the use of hops in "beer," declaring hops a "wicked weed." The ban reflected both a cultural preference for traditional "ales" and the influence of those who profited from gruit sales. However, such bans were eventually lifted as tastes and economic considerations changed.

One of the most significant milestones in brewing history was the enactment of the Bavarian Purity Law—Reinheitsgebot (ryne-HYTS-guh-boht)—on April 23, 1516. Introduced by Duke Wilhelm IV and his brother Duke Ludwig X of Bavaria, this law stipulated that beer could only be made from water, barley, and hops. At the time, yeast was not mentioned in the Reinheitsgebot because its role in fermentation was not yet understood; the existence of microorganisms like yeast was not discovered until the 17th century, and their role in fermentation was elucidated by Louis Pasteur in the 19th century.

The Reinheitsgebot aimed to standardize beer quality, prevent the use of harmful additives, and reserve wheat and rye for bread making, ensuring adequate supplies of these grains for food. This law played a significant role in popularizing hops as an essential ingredient in beer across Bavaria and later throughout Germany. It also had political and economic implications: by mandating specific ingredients, it controlled the production of beer, influencing grain markets and limiting competition from brewers who used alternative ingredients. The law reinforced the power of the state over brewing practices and reduced the influence of the Church, which had previously controlled gruit production.

Despite the advantages of hops, their adoption faced resistance. Economic interests profiting from gruit sales opposed the shift to hops, sometimes influencing local laws to restrict their use. Traditionalists and some consumers preferred the flavors of gruit ales and resisted change. There were also unfounded fears that hops were harmful or had sedative effects.

Role of Women in Medieval Europe

Outside the monasteries, brewing in medieval Europe was predominantly a domestic and female-led activity. Women, known as alewives or brewsters, were the primary producers of ale—a vital staple in the medieval diet. They brewed ale for household consumption and often sold any surplus to neighbors and townsfolk, making brewing an important source of income for many families.

Alewives were common and essential figures in towns and villages across Europe. They were often identifiable by distinctive signs displayed outside their homes or establishments:

- Ale Stake: A horizontal pole or rod projecting from the building, sometimes adorned with evergreen boughs or other symbols, signaling that ale was available within.
- Broomstick: A broom or broomstick hung above the door served as a practical advertisement, indicating that fresh ale was ready for sale.

These visual markers were crucial in a largely illiterate society, guiding customers to where they could purchase ale.

Intriguingly, some historians suggest that the stereotypical imagery associated with witches—women wearing tall, pointed hats, stirring cauldrons, accompanied by cats—may have roots in the medieval alewife:

- Tall, Pointed Hats: Alewives often wore tall hats to stand out in crowded marketplaces, making themselves easily identifiable to customers seeking ale amid the bustle of market days.

- Cauldrons: Large cauldrons were essential tools for brewing, used to boil the wort during the brewing process. The image of a woman stirring a cauldron became emblematic.
- Cats: Alewives kept cats to protect their grain stores from rodents. Grain was a valuable commodity, and cats served as natural pest control.
- Broomsticks: The broomstick, besides being an advertising tool, was a common household item, symbolizing domesticity and, later, magical flight in folklore.

These associations may have contributed to the later iconography of witches in folklore and popular culture. While the connection is subject to debate among historians, it offers a compelling lens through which to view the social dynamics of the time.

As brewing became more commercialized and profitable from the 14th century onward, women were gradually pushed out of the industry. Several factors contributed to this shift:

- Guild Formation and Legal Restrictions: Brewing guilds began to form, and membership was often restricted to men. These guilds regulated the production and sale of ale and beer, controlling prices, quality, and who could legally brew for commercial purposes. Women found themselves excluded from these professional organizations.
- Economic Competition: Men entering the brewing trade sought to dominate the market. By controlling resources and access to markets, they undermined the economic independence of female brewers. Women's brewing was increasingly legislated against or taxed heavily.
- Social Stigmatization and Misogynistic Stereotypes: Alewives faced growing social scrutiny. Negative stereotypes portrayed them as dishonest, unclean, or immoral. Satirical literature and

popular songs of the time often depicted alewives engaging in deceitful practices, damaging their reputations.

- Accusations of Witchcraft: During periods of social tension and the witch hunts of the early modern period, women's association with brewing equipment—cauldrons, broomsticks, and cats—made them targets for accusations of witchcraft. Claims of using magic or sorcery in their brewing practices were sometimes employed to discredit female brewers and justify their removal from the trade.

The combination of legal barriers, economic pressures, and social stigmatization led to the decline of women's prominence in brewing. By the 16th century, brewing had largely become a male-dominated profession in many parts of Europe. Women were systematically excluded from an industry they had nurtured and sustained for centuries.

This transition not only affected the livelihoods of countless women but also altered the social fabric of communities. Brewing had provided women with economic independence and a degree of social standing. Its loss meant diminished roles in public life and reduced opportunities for financial self-sufficiency.

The story of alewives and brewsters highlights the complex interplay of gender, economics, and societal norms in medieval Europe. It reflects how women's contributions to essential industries were significant yet vulnerable to shifting power dynamics and cultural narratives.

Understanding this history offers valuable insights into the broader patterns of women's labor and the social forces that have shaped gender roles over time. It also underscores the importance of recognizing and preserving the legacy of women's work in brewing and other crafts that have been historically overlooked or marginalized.

The Discovery of Yeast

While yeast had been used in brewing for millennia, its existence and role in fermentation were not scientifically understood until much later. In the 17th century, Dutch scientist Antonie van Leeuwenhoek (1632–1723) used one of the earliest microscopes he developed to observe fermenting beer. In 1680, he described seeing tiny, moving organisms in the liquid, referring to them as "animalcules" or "little animals." Van Leeuwenhoek wrote to the Royal Society of London about these observations, noting:

> *I have observed some very small animalcules, very diligently moving themselves.*

Despite observing these microorganisms, he did not comprehend their role in fermentation. Often referred to as the "Father of Microbiology" for his pioneering work in microscopy, van Leeuwenhoek's discovery of these living creatures was a groundbreaking step. However, the connection between yeast and fermentation would remain elusive for another two centuries.

It wasn't until the 19th century, over 200 years after van Leeuwenhoek's discovery, that the role of yeast in fermentation was fully elucidated by the French chemist and microbiologist Louis Pasteur (1822–1895). In 1857, Pasteur began studying fermentation at the request of the wine industry, which was facing spoilage issues. Through meticulous experiments, he demonstrated that fermentation was a biological process carried out by living organisms—yeast.

In 1861, Pasteur conducted experiments that falsified—disproved—the hypothesis of spontaneous generation, showing that microorganisms come from other microorganisms rather than from non-living matter. His approach followed the six steps of the scientific method. First, (1) he observed that sterilized liquids exposed to air developed microbial growth, while sealed containers did not. This led him to (2) formulate the hypothesis that microorganisms do not arise spontaneously but come from preexisting organisms in the air. To test this, (3) he designed the famous swan-neck flask experiment, placing a broth in flasks with curved necks that prevented airborne particles from reaching the liquid. He boiled the broth to sterilize it, leaving the flasks exposed to air but protected from outside airborne particles by the flask's curved neck. The broth remained sterile as long as no particles from the outside air reached it, but once the flasks were tilted or broken, allowing airborne particles to enter, microbial growth quickly appeared. (4) After analyzing the data, Pasteur concluded that he had falsified the hypothesis of spontaneous generation and found that the hypothesis that life was introduced from external sources was plausible. His experiment (5) was replicated by other scientists in different locations, and (6) his publication was peer-reviewed, providing further credibly to his findings. This application of the scientific method conclusively falsified the long-held theory of spontaneous generation.

In 1876, Pasteur published *Études sur la Bière* (Studies on Beer), where he detailed his findings on fermentation and the roles of yeast and bacteria in brewing. He introduced the concept of pasteurization, a process of heating liquids to kill unwanted microorganisms, which helped preserve beer and wine by reducing spoilage and extending shelf life. Pasteur's contributions laid the foundation for modern microbiology, which is critical for understanding and controlling fermentation in brewing.

The Industrial Revolution and the Rise of Lagers

The Industrial Revolution of the 18th and 19th centuries brought about transformative changes in brewing. Technological advancements, such as the development of the steam engine by James Watt in 1765, revolutionized beer production. Breweries could now operate on a larger scale, increasing their output and efficiency. The introduction of mechanized equipment allowed for more precise control over the brewing process, enhancing consistency and quality.

One of the most significant developments during this period was the rise of lager beer. Originating in Bavaria in the 15th century, lagering involved storing beer in cold caves or cellars to ferment and mature at lower temperatures. However, it wasn't until the 19th century that lagers began to dominate the beer market, largely due to technological advancements.

The invention of refrigeration by Carl von Linde in the 1870s allowed for the precise control of fermentation temperatures necessary for lager brewing. Lagers require cooler fermentation conditions (typically between 7°C and 13°C or 45°F and 55°F), which were difficult to maintain year-round before artificial refrigeration. With this technology, brewers could produce lagers consistently, regardless of the season.

The development of pale malt through innovations in kilning techniques led to the production of lighter-colored beers. Traditional malting involved drying malt over direct heat from wood or coal fires, which often imparted darker colors and smoky flavors to the malt. In the early 19th century, the invention of indirect kilning methods allowed maltsters to control drying temperatures more precisely and avoid direct contact with smoke, resulting in paler malts.

In 1842, the first Pilsner lager was brewed in the city of Plzeň (Pilsen) in what is now the Czech Republic, by brewer Josef Groll. This beer was clear, golden, and crisp, contrasting with the darker, heavier ales and lagers of the time. The Pilsner style quickly gained popularity across Europe and eventually worldwide.

Brewing lighter beers like Pilsner presented unique challenges. The delicate flavors and pale appearance meant that any imperfections or off-flavors were highly noticeable, unlike in darker ales and beers where robust malt profiles and stronger gruit and hop flavors mask minor defects. Consistency and purity in Pilsner lager became paramount, requiring precise control over ingredients and brewing conditions.

Technological advancements made it possible to overcome these challenges:

- Hydrometer: Introduced in the late 18th century, the hydrometer allowed brewers to measure the sugar content of the wort and monitor fermentation progress accurately. This tool enabled better control over the brewing process and ensured desired alcohol levels and flavor profiles were achieved.
- Thermometer: The widespread use of thermometers provided brewers with the ability to maintain exact temperatures during mashing and fermentation. Precise temperature control is crucial for enzyme activity during mashing and for yeast performance during fermentation, directly affecting the beer's flavor and clarity.
- Advancements in Microbiology: The work of scientists like Louis Pasteur improved the understanding of yeast and fermentation. Recognizing yeast as a living organism that could be managed and cultivated led to the isolation of pure yeast strains. In 1883, Emil Christian Hansen at the Carlsberg Brewery developed methods to propagate pure yeast cultures, reducing contamination and improving consistency.
- Water Chemistry: Understanding the importance of water composition, brewers began treating water to replicate the soft water of Plzeň, which was ideal for producing light lagers. Adjusting mineral content helped in achieving the desired beer characteristics and stability.
- Filtration and Clarification Techniques: Advances in filtration methods allowed for the removal of yeast and particulate matter, resulting in the brilliant clarity that became a hallmark of lagers like Pilsner.

These technological innovations were essential in producing high-quality lighter beers on a large scale. The ability to control every aspect of the brewing process minimized the risk of defects that would be apparent in such delicate beers.

The industrialization of brewing, combined with the growing preference for clear, crisp, and light lagers, shifted consumer tastes. Lagers became the dominant beer style, favored for their refreshing qualities and consistency. Large breweries capitalized on this trend, utilizing economies of scale and advanced technology to mass-produce lager beers.

This period marked a significant transformation in the brewing industry, where science and technology played pivotal roles in shaping beer production. The challenges of brewing lighter beers spurred innovation, leading to practices and equipment that set new standards for quality and efficiency. The success of lagers during the Industrial Revolution underscores how technological advancements enabled brewers to meet consumer demands for cleaner, more consistent beers, ultimately changing the global beer landscape.

Ales in Britain and the Emergence of the India Pale Ale

While lagers were gaining popularity on the European continent, Britain remained steadfast in its ale brewing traditions. The Industrial Revolution influenced British brewing by introducing technological advancements that improved efficiency and consistency, but the British palate continued to favor ales over lagers. Innovations in malting, milling, and brewing equipment allowed British brewers to refine their ales, leading to the development of new styles and improved quality.

During this era, the brewing industry in Britain saw the rise of the India Pale Ale (IPA), a beer style that has become iconic. The popular narrative suggests that IPAs were developed by English brewers in the late 18th century to survive the long sea voyage to India. According to this story, brewers increased the alcohol content and hopping rates of their pale ales to preserve the beer during the months-long journey, as hops and alcohol have natural preservative properties.

However, modern beer historians have scrutinized this tale. George Hodgson, a brewer at the Bow Brewery in London, is often credited with popularizing beer exports to India around 1780. His brewery was conveniently located near the East India Company's docks, facilitating trade. Hodgson's October beer, a well-hopped pale ale intended for aging, was shipped to India and became popular among British expatriates.

While it is true that heavily hopped beers were sent to India and that hops aided in preservation, the idea that the IPA was specifically invented for that purpose is likely an oversimplification. Beers of various styles, including porters and stouts, were successfully shipped to India without spoilage. The term "India Pale Ale" began appearing in advertisements and brewery records in the 1830s, suggesting that the style evolved over time in response to consumer demand rather than being a singular invention.

While the IPA did become associated with exports to India, the romanticized story of its invention to withstand long sea voyages is more legend than fact. The reality is more nuanced, involving gradual stylistic developments, market demands, and the natural evolution of brewing practices.

Prohibition and the Decline of Breweries

In the United States, the early 20th century brought significant challenges to the brewing industry, culminating in the enactment of Prohibition. The temperance movement, which gained momentum throughout the 19th century, was driven by social concerns about the negative effects of alcohol on individuals, families, and society. Women played a pivotal role in this movement, with organizations like the Women's Christian Temperance Union (WCTU) advocating for the reduction or elimination of alcohol consumption to combat domestic abuse, financial instability, and moral degradation.

The marginalization of women from the brewing industry, a process that began centuries earlier in Europe, had shifted alcohol production from a domestic, often female-led activity to a commercial, male-dominated enterprise. As brewing became increasingly detached from the home, alcohol consumption grew more associated with male-dominated spaces like saloons, which were viewed as centers of vice, corruption, and excessive drinking. The absence of women's influence in brewing and alcohol consumption may have contributed to shifting societal attitudes, reinforcing the perception of alcohol as a destructive force rather than a communal or familial staple.

These changing attitudes were among the factors that fueled the temperance movement. Advocates believed that alcohol was responsible for many social ills, including poverty, crime, and the breakdown of family structures. The belief that men were spending excessive amounts of money in bars, leading to dysfunctional families and communities, was a significant motivator for many Prohibition supporters.

The efforts of the temperance movement culminated in the ratification of the 18th Amendment and the passage of the Volstead Act in 1919, which took effect on January 17, 1920. Prohibition made the manufacture, sale, and transportation of alcoholic beverages illegal nationwide.

Before Prohibition, there were over 1,500 breweries operating in the United States, producing a diverse range of beer styles. The ban on alcohol devastated the brewing industry. While some breweries attempted to survive by producing non-alcoholic beverages like sodas and near beer (beer with less than 0.5% alcohol by volume), many were forced to close. By the time Prohibition was repealed with the 21st Amendment on December 5, 1933, the number of breweries had dwindled dramatically.

After Prohibition, only 756 breweries resumed operations. The industry faced additional challenges, including the Great Depression, World War II, and restrictive regulations. Economic pressures led to consolidation within the industry, with larger breweries buying out smaller ones. By the 1980s, the number of breweries in the United States had declined to fewer than 100, and the market was dominated by a few large corporations producing mainly light, mass-produced lagers.

The legacy of Prohibition had long-lasting effects on the American brewing landscape. The social and cultural factors that contributed to its enactment—including the earlier marginalization of women from brewing—highlight the complex interplay between gender dynamics, societal attitudes toward alcohol, and regulatory policies. While it's speculative to assert that continuous female involvement in brewing would have prevented Prohibition, the shifts in brewing practices and perceptions of alcohol consumption did play a role in shaping the conditions that led to this significant period in history.

Understanding this context provides valuable insight into how economic interests, social movements, and gender roles have historically influenced brewing and alcohol consumption. It underscores the importance of inclusivity and diverse perspectives in shaping industries and societal attitudes. As the brewing industry continues to evolve, recognizing and learning from these historical patterns can inform more equitable and sustainable practices for the future.

The Modern Era and the Craft Beer Movement

The late 20th century marked the beginning of the craft beer movement, a resurgence of small-scale brewing that celebrated diversity, innovation, and traditional methods. This movement was driven by a desire to challenge the dominance of large commercial breweries and to revive the art of brewing unique, flavorful beers

—particularly ales, which had become less common in the U.S. market.

A pivotal moment came in 1978, when President Jimmy Carter signed H.R. 1337, a bill that included an amendment legalizing homebrewing at the federal level, effective February 1, 1979. Prior to this, homebrewing had been illegal since Prohibition, even though Prohibition had been repealed in 1933. This change in legislation opened the door for hobbyists and entrepreneurs to explore brewing on a small scale.

Homebrewers began experimenting with a variety of beer styles, including those not widely available from commercial brewers. They focused on ales, which are generally quicker and easier to produce on a small scale compared to lagers. The first modern microbrewery in the United States, New Albion Brewing Company, was founded by Jack McAuliffein 1976 in Sonoma, California. Although New Albion closed in 1982 due to financial difficulties, it inspired others to follow.

Throughout the 1980s and 1990s, the number of microbreweries and brewpubs grew rapidly. These craft brewers emphasized quality, flavor, and traditional brewing methods. They reintroduced a variety of beer styles, including pale ales, stouts, porters, and IPAs, catering to consumers seeking alternatives to mass-produced lagers.

By the early 2000s, the craft beer movement had gained significant momentum. According to the Brewers Association, the number of breweries in the United States grew from around 1,500 in 2007 to over 9,000 by 2021. This remarkable resurgence reflects a renewed interest in diverse beer styles and locally produced beverages.

Despite the legalization, the craft beer movement faced resistance similar to that encountered in earlier centuries. Large commercial breweries wielded significant influence over distribution networks, retail spaces, and marketing channels. Craft brewers often struggled to access these avenues, facing challenges in bringing their products to market. Additionally, regulatory hurdles at the state and local levels sometimes impeded the growth of small breweries. For example, laws in some states limited the alcohol content of beer or restricted the ability of breweries to sell directly to consumers.

Just as medieval brewers contended with monopolies and economic control exercised by the Church and local authorities through the gruitrecht, modern craft brewers have navigated a landscape where economic power can limit access and opportunity. The craft beer movement, like the transition from gruit to hops, represents a shift driven by consumer preference, innovation, and a desire for diversity in beer.

The success of craft brewers caught the attention of large brewing companies. In response to market demand, big brewers began producing their own versions of craft-style beers and acquiring successful craft breweries. This trend mirrors historical patterns where dominant entities adopt innovations initially introduced by smaller competitors to maintain market share.

The Community of Beer Enthusiasts and Educational Advancements

The growth of the brewing industry has been accompanied by the establishment of numerous brewing organizations, certification programs, and educational institutions dedicated to advancing the art and science of brewing.

- Brewers Association (BA): Founded in 2005 through the merger of the Association of Brewers and the Brewers' Association of America, the BA represents small and independent American craft brewers. It provides resources, advocacy, and events like the Great American Beer Festival, promoting the interests of craft brewers nationwide.
- Beer Judge Certification Program (BJCP): Established in 1985, the BJCP certifies and ranks beer judges through an examination and monitoring process. It provides style

guidelines used worldwide in homebrew competitions, fostering a standardized understanding of beer styles and quality assessment.

- Cicerone Certification Program: Launched in 2007 by Ray Daniels, this program offers professional credentials for beer industry professionals, focusing on beer styles, storage, service, and pairing. It aims to elevate the knowledge and service standards in the hospitality industry.
- American Society of Brewing Chemists (ASBC) and Master Brewers Association of the Americas (MBAA): These organizations support brewing science and technology, providing research, publications, and networking opportunities for brewing professionals.
- Educational Institutions: Universities and colleges have developed brewing science programs, such as the Siebel Institute of Technology in Chicago, University of California, Davis, and Oregon State University. These programs offer degrees and certificates in brewing science, fermentation, and related fields, contributing to the professionalization of the industry.

The brewing community is further enriched by homebrewing clubs, beer festivals, and competitions. Enthusiasts gather to share knowledge, taste new creations, and celebrate the diversity of beer. Events like the National Homebrew Competition, organized by the American Homebrewers Association, encourage experimentation and innovation among hobbyists.

Role of Women in the Modern Era

In the modern era, the craft beer movement has witnessed a significant re-emergence of women in brewing roles, marking a resurgence of their historical contributions to the industry. After centuries of marginalization, women are reclaiming their place as brewmasters, brewery owners, educators, and industry leaders, bringing fresh perspectives and innovations to the world of beer.

As the craft beer movement gained momentum in the late 20th and early 21st centuries, more women began entering the brewing industry. This resurgence aligns with broader social movements advocating for gender equality and challenging traditional gender roles. Women are drawn to brewing not only for the opportunity to create unique beers but also to help reshape an industry that had become predominantly male.

Several organizations have been instrumental in supporting and promoting women in the brewing industry:

- Pink Boots Society (founded in 2007 by Teri Fahrendorf): A global nonprofit dedicated to assisting, inspiring, and encouraging women in the brewing profession through education. The society offers scholarships, networking opportunities, and resources to help women advance their careers. They also collaborate with the International Women's Collaboration Brew Day (IWCBD), established in 2014 by Sophie de Ronde in the UK, where women brewers worldwide create beers under a common theme to raise funds for educational scholarships.
- Women of the Vine & Spirits (founded in 2015 by Deborah Brenner): An organization that provides mentorship programs, networking events, and educational resources, connecting women across all sectors of the alcohol beverage industry to promote inclusion and equality.

Several women have made significant contributions to the modern brewing industry:

- Carol Stoudt (founded Stoudts Brewing Company in 1987): Becoming one of the first female brewmasters and brewery owners in the United States since Prohibition, she is renowned for producing high-quality German-style lagers and ales. After 33 years, she retired and closed Stoudts Brewing Company in 2020, leaving a lasting legacy as a trailblazer for women in brewing.

- Kim Jordan (co-founded New Belgium Brewing Company in 1991): Under her leadership as CEO until 2015, New Belgium grew into one of the largest and most respected craft breweries in the United States, known for its innovative beers, environmental sustainability efforts, and employee ownership model. She continues to serve as Executive Chair of the Board.
- Megan Parisi: An accomplished brewer who has worked at several prominent breweries, including Wormtown Brewery, Cambridge Brewing Company, Bluejacket Brewery, and currently as the head brewer at the Samuel Adams Boston Brewery. She has contributed to the development of new beer styles and brewing techniques, including barrel-aged and sour beers.
- Celeste Beatty (founded Harlem Brewing Company in 2000): Recognized as one of the first African American women to own a brewery in the United States, she focuses on community engagement and celebrating African heritage in brewing. Her beers often incorporate ingredients from Africa and the African diaspora.
- Tara Nurin: An award-winning beer writer and educator, author of "A Woman's Place Is in the Brewhouse: A Forgotten History of Alewives, Brewsters, Witches, and CEOs" (published in 2021), which highlights the historical and contemporary contributions of women in brewing.

Despite their growing presence, women remain underrepresented, especially in production and leadership roles. According to the Brewers Association's 2021 Brewery Operations Benchmarking Survey:

- Women make up approximately 23.7% of the overall brewery workforce.
- Only about 7.5% are involved in production roles, such as brewers and production staff.

Women in brewing often face implicit bias and stereotypes questioning their expertise. Instances of discrimination and harassment have highlighted the need for cultural change within the industry. In 2021, the brewing industry faced a reckoning as numerous women shared their experiences of sexism and harassment, leading to increased calls for inclusivity and safer work environments.

The increasing involvement of women has enriched the brewing industry. Women brewers are:

- Introducing new flavors and techniques: Expanding the diversity of beer offerings by experimenting with novel ingredients and brewing methods.
- Prioritizing sustainability and ethics: Leading initiatives in environmental sustainability, community engagement, and ethical business practices.
- Challenging marketing norms: Helping breweries move away from gendered stereotypes in marketing, appealing to a broader and more inclusive audience.

The re-emergence of women in brewing is a global phenomenon, with notable contributions worldwide:

- Africa: Women have traditionally brewed beer as part of cultural practices. Modern initiatives, such as those led by Ngozi Nwofor of Nigeria's Wavecrest College and South Africa's Dope Beer Hacks, support women brewers and entrepreneurs, promoting economic empowerment.
- Asia: In Japan, brewers like Miki Sawada of Kyoto Brewing Company are making their mark. In China, Liu Fang co-founded Bravo Brewing Co., contributing to the country's growing craft beer scene.
- Latin America: Women brewers in Brazil, Argentina, and Mexico are gaining recognition. Organizations such

as Mujeres Cerveceras de Argentina provide support and advocacy, fostering a community among female brewers.

- Europe: In the UK, Dea Latis (founded in 2010) promotes beer to women and supports women in the industry. Katharina Kurz, co-founder of BRLO in Germany, has been influential in Berlin's craft beer movement. In the Netherlands, Henriette Nikken became Heineken's first female brewmaster in 2017.

Access to education and professional development has been crucial. Institutions like the Siebel Institute of Technology, University of California, Davis, and Oregon State University have seen increased enrollment of women in their brewing programs. Scholarships and mentorship programs, such as those offered by the Pink Boots Society, facilitate career advancement and knowledge sharing.

As women continue to reclaim their historical role in brewing, they enrich the industry with creativity, diversity, and a commitment to inclusivity. Their contributions challenge traditional norms and open new possibilities for the art and science of brewing. The ongoing efforts to promote gender equality and address systemic issues within the industry are essential for fostering an environment where all brewers can thrive.

Tradition, Innovations, and Future Trends

As the brewing community has become more diverse and inclusive, so also have the technological innovations that shape the industry. Automated brewing systems streamline production, allowing for precise control over every aspect of the brewing process. Advanced fermentation control technologies enable brewers to maintain optimal conditions for yeast activity, enhancing the consistency and quality of the final product. Sophisticated quality assurance processes, including lab testing and sensory evaluation, ensure that each batch of beer meets the highest standards of style, flavor, and safety.

The continuous interplay of rationalism, empiricism, and the scientific method has driven the evolution of brewing practices. As we look to the future, the ability to blend tradition with innovation will be essential. Brewers who cultivate strong analytical skills and embrace a mindset of continuous learning will be well-positioned to contribute to the ongoing advancement of the art and science of brewing.

Conclusion

Brewing is, in many ways, a modern form of alchemy. It embodies the artful and scientific transformation of simple ingredients—water, malt, hops, and yeast—into the golden elixir we call beer. Just as historical alchemists sought to turn base materials into gold, brewers blend precise chemical processes with creative techniques to craft unique and flavorful brews. This transformative journey is not merely about combining ingredients; it's about harnessing the principles of rationalism and empiricism, and applying the scientific method to achieve extraordinary results.

Throughout history, brewers have built upon centuries of accumulated knowledge, combining traditional practices with cutting-edge science. The integration of modern technology into brewing represents a continuation of this scientific pursuit, where observation, experimentation, and analysis drive innovation. Today's brewers stand on the shoulders of those ancient Sumerian brewsters and medieval artisans, honoring the past while embracing the future.

This fusion of historical wisdom and modern ingenuity requires not just technical proficiency but also a proactive approach to problem-solving and a willingness to question and refine existing methods. By adopting both deductive reasoning (rationalism) and inductive reasoning (empiricism), brewers implement the scientific method to enhance their craft. This critical thinking enables them to push the boundaries of what is possible, transforming simple ingredients into extraordinary creations.

As we embark on this journey through the art and science of brewing, this book aims to provide you with the foundational knowledge and structured approaches necessary to become a critical-thinking brewmaster. By intertwining historical developments with philosophical and methodological frameworks, we not only trace how rationalism, empiricism, and the scientific method have advanced brewing but also actively engage you in developing the skills essential for innovation and problem-solving.

The journey of beer is a testament to human creativity, resilience, and the power of thoughtful inquiry. From ancient civilizations to the modern craft beer movement—including the re-emergence of women in brewing—the evolution of beer reflects our unending quest to transform base materials into something golden. As you delve deeper into the chapters ahead, you'll be empowered to blend tradition with innovation, apply logical reasoning and empirical evidence to your brewing practices, and contribute to the ongoing evolution of this timeless craft.

By fostering a spirit of exploration, curiosity, and a commitment to continual improvement, you will join the ranks of those who have dared to push the boundaries of brewing. The art of brewing remains vibrant and dynamic, ever adapting to new challenges and opportunities. With each batch you brew, you participate in this modern alchemy, transforming simple ingredients into something truly remarkable. Together, we continue this journey, honoring the past while pioneering the future of beer.

> *Brewing, at its best, is not merely a technical achievement but a moral one—an act of stewardship that honors the labor of those forgotten, the wisdom of those passed, and the promise of those yet to come. To brew with integrity is to engage in a quiet ethic: to remember, to refine, and to refuse the erasure of those who shaped the craft before us.*

**

Review Questions

True/False Questions

1. True or False: The Sumerians brewed beer as early as 5,000 BCE and composed the "*Hymn to Ninkasi*," which served as a prayer and a brewing recipe.

2. True or False: Beer was only consumed by the lower classes in ancient Egypt and not by the pharaohs or higher social classes.

3. True or False: The Reinheitsgebot (ryne-HYTS-guh-boht), or Bavarian Purity Law of 1516, stipulated that beer could only be made from barley, water, and hops.

4. True or False: Louis Pasteur's research on yeast and fermentation led to the development of pasteurization, which helped preserve beer and reduce spoilage.

5. True or False: The craft beer movement of the late 20th century aimed to challenge the dominance of large commercial breweries by celebrating diversity, innovation, and traditional brewing methods.

Multiple Choice Questions

6. Which ancient civilization is known for composing the *"Hymn to Ninkasi*," a text that highlights their empirical knowledge of brewing?
A) Greeks
B) Egyptians
C) Romans
D) Sumerians

7. The Reinheitsgebot, enacted in 1516, aimed to ensure the purity and quality of beer by allowing which ingredients?
A) Barley, wheat, hops, and water
B) Barley, water, hops, and yeast
C) Barley, water, yeast, and spices
D) Barley, water, hops

Brewer Vignettes

8. Brewer Vignette 1: Imagine you are a brewer in ancient Mesopotamia, around 5,000 BCE. You follow the *"Hymn to Ninkasi*" as your brewing guide. Describe the significance of this hymn in your daily brewing practices and how it reflects your empirical observations and experiments with ingredients and processes.
A) The hymn was a mythological story with no relevance to brewing practices.
B) The hymn focused on agricultural practices rather than brewing.
C) The hymn was used solely as a religious chant with no practical application to brewing.
D) The hymn detailed specific steps and ingredients used in brewing, reflecting empirical observations and experiments.

9. Brewer Vignette 2: As a monk in a medieval European monastery, you are responsible for brewing beer according to the Rule of St. Benedict. Explain how your brewing practices are influenced by rationalistic top down deductive reasoning, and how your systematic approach ensures the production of high-quality beer for the community.
A) Brewing was a secondary activity with minimal importance.
B) Rational planning and theoretical understanding ensured consistent high-quality beer production.
C) The monks used random methods without any systematic approach.
D) The Rule of St. Benedict discouraged brewing practices entirely.

10. Brewer Vignette 3: You are a craft brewer in the late 20th century, inspired by the resurgence of small-scale brewing. Describe how you combine traditional knowledge with experimental approaches to create new and unique beer styles. Highlight the role of empiricism and the scientific method in your brewing innovations.
A) Traditional knowledge was combined with empirical experimentation to create innovative beers.
B) Experimental approaches were considered unnecessary and avoided.
C) The movement focused only on replicating historical recipes without innovation.
D) The craft beer movement rejected traditional knowledge entirely.

**

Correct Answers
True/False Questions

1. True
2. False (Beer was enjoyed by people of all social classes in ancient Egypt.)
3. True
4. True
5. True

Multiple Choice Questions
6. D) Sumerians
7. D) Barley, water, hops
Brewer Vignettes
8. D) The hymn detailed specific steps and ingredients used in brewing, reflecting empirical observations and experiments.
9. B) Rational planning and theoretical understanding ensured consistent high-quality beer production.
10. A) Traditional knowledge was combined with empirical experimentation to create innovative beers.

**

Beyond The Chapter

- Bamforth, C. W. (2003). Beer: Tap into the Art and Science of Brewing. Oxford University Press.
  - *Provides an in-depth look at the historical context and evolution of brewing techniques and practices.*
- Hornsey, I. S. (2003). A History of Beer and Brewing. Royal Society of Chemistry.
  - *Explores the comprehensive history of beer and brewing from ancient civilizations to modern times.*
- Pasteur, L. (1876). Études sur la Bière. Paris: Gauthier-Villars.
  - *Discusses Pasteur's groundbreaking work on fermentation and its impact on brewing.*
- Oliver, G. (2011). The Oxford Companion to Beer. Oxford University Press.
  - *Offers a broad range of topics related to beer, including historical aspects and brewing traditions.*
- Calagione, S. (2011). Extreme Brewing: An Enthusiast's Guide to Brewing Craft Beer at Home. Quarry Books.
  - *Presents the evolution and influence of craft brewing, focusing on innovative brewing practices.*
- Daniels, R. (1996). Designing Great Beers: The Ultimate Guide to Brewing Classic Beer Styles. Brewers Publications.
  - *Examines the historical development of various beer styles and their traditional brewing methods.*

Weblinks

- Ancient Origins of Brewing - History Channel: https://www.history.com/search?q=Beer
- Beer in Ancient Egypt - Archaeology Magazine: https://archaeology.org/?s=Beer

- Sumerians and Beer - Smithsonian Magazine: https://www.smithsonianmag.com/history/ancient-sumerians-enjoyed-a-rich-variety-of-beers-180970116/
- Medieval Monastic Brewing - Brewminate: https://brewminate.com/?s=Medieval
- The Reinheitsgebot - Bavarian Purity Law: https://bavaria.travel/gourmet-stories/bavarian-beer/?seed=1720645402326
- Louis Pasteur's Impact on Brewing - Science History Institute: https://www.sciencehistory.org/?s=Louis+Pasteur
- The Craft Beer Movement - Brewers Association: https://www.brewersassociation.org/search/?_sf_s=History
- Modern Brewing Technology - Craft Beer & Brewing Magazine: https://beerandbrewing.com/search/?search=Technology
- History of Brewing Techniques - Brew Your Own: https://byo.com/?s=History&post_type=any&orderby=relevance

These references provide a deeper dive into the history and science of brewing, offering valuable insights for both homebrewers and professionals.

***

Chapter 2: The Symphony of Malt, Adjuncts, and Sugars

From grain to glass, the harmony of malt, adjuncts, and sugars composes a symphony of flavors.

Imagine walking into a bustling brewery, the air filled with the rich aroma of grains and the subtle sweetness of fermenting sugars. This sensory experience starts with the fundamental ingredients of brewing: malt, adjuncts, and sugars. Picture malted barley being transformed through processes of germination and kilning, and adjuncts like corn and rice adding unique flavors and textures. Understanding their roles and transformations is crucial for any brewer, from the novice homebrewer to the seasoned professional.

In brewing, the selection of malt, adjuncts, and sugars plays a critical role in determining the flavor, aroma, color, and overall quality of the beer. Each ingredient brings unique properties that influence the brewing process and the final product. Understanding these components allows brewers to craft beers with precise characteristics and consistent quality. This chapter delves into the intricacies of malt, adjuncts, and sugars, exploring their historical context, chemical properties, and practical applications in brewing.

Malt

The Foundation of Beer

Malt, primarily derived from barley, is the cornerstone of beer production. Barley or other grains are germinated and kilned, forming the base ingredient in beer brewing. The malting process modifies the grain's endosperm—the starchy interior of the grain—and activates enzymes essential for brewing.

The modification of the endosperm correlates with the growth of the acrospire, the sprout of a germinating barley grain. The length of the acrospire indicates the level of modification during the malting process, which determines the grain's crushability and the starch's convertibility into sugars by enzymes formed during germination.

The Malting Process

The process of malting begins with steeping, where barley grains are soaked in water to initiate germination. This hydration activates enzymes and prepares the grains for the next stage. After steeping, the grain is allowed to germinate, during which enzymes like amylases develop. These enzymes catalyze biochemical reactions, breaking down the starches in the endosperm into simpler sugars.

Once the acrospire grows to a desired length, indicating sufficient modification, the germination process is halted by drying the grain in a kiln—a process known as kilning. Kilning not only dries the malt but also develops its flavor and color. At higher kiln temperatures, the malt's diastatic strength decreases while its color darkens and caramelization flavors intensify. Conversely, at lower kiln temperatures, the malt retains greater diastatic strength, resulting in a lighter color and milder caramelization flavors.

Key Factors in Malt Production

1. Diastatic Power: Malt's diastatic power is a measurement of the malt's enzyme content, indicating its ability to convert starches into fermentable sugars. It is measured in degrees Lintner. High diastatic power is essential for efficient starch conversion, especially when using adjuncts like rice or corn, which lack their own enzymes. Malts with high diastatic power can convert not only their own starches but also those of adjuncts, making them versatile in various brewing applications.
2. Nitrogen and Protein Content: Nitrogen is an essential element in brewing found in proteins. By measuring the nitrogen content, brewers can estimate the total protein content in malt using the formula: Total Protein Content (%) = Nitrogen Content (%) × 6.25. Higher modification of malt increases

soluble nitrogen, improving protein modification and reducing haze in beer. Well-modified malts have undergone thorough malting, breaking down internal grain structures to convert proteins and starches into soluble forms. This makes them more accessible during mashing, improving extract efficiency and enzyme activity. Lower total nitrogen content generally leads to clearer beer with less risk of haze formation.

3. Color: The degree of kilning affects the malt's color, measured in degrees Lovibond (°L) or European Brewing Convention (EBC) units. Lovibond is a scale where higher values indicate darker colors, while EBC units measure the color of beer and malt, ranging from pale yellow to dark brown. Kilning at higher temperatures produces darker malts with more complex flavors but reduces enzymatic activity. The color of the malt influences the final hue of the beer, ranging from pale gold to deep brown. Brewers carefully select malts based on the desired color and flavor profile of their beer.

Types of Pale Malt

1. Six-Row Brewers' Malt: Six-row barley has higher enzyme and protein content compared to two-row barley, making it suitable for adjunct-heavy brews. It has a thicker husk, which contributes to higher tannin levels and aids in efficient sparging but can cause haze and astringency if not managed properly. Six-row malt is often used in American lagers, where adjuncts like corn and rice are added to lighten the body and flavor. The high diastatic power of six-row malt ensures the complete conversion of starches from both the malt and adjuncts.

2. Two-Row Brewers' Malt: Two-row barley is preferred for its lower protein and tannin content, resulting in clearer, less astringent beers. It has a thinner husk and provides a cleaner malt profile, making it ideal for various beer styles. Two-row malt is the standard choice for many craft brewers due to its balance of flavor, enzymatic activity, and clarity. It forms the

backbone of a wide range of beer styles, from pale ales to stouts.

3. Pale Ale Malt: The standard British malt, pale ale malt, undergoes a long kilning process, resulting in rich, biscuity flavors. Its lower enzymatic activity necessitates careful mashing but contributes to the malt-forward character of traditional British ales. Pale ale malt is known for its ability to produce complex, flavorful beers with a distinctive malt character. It is often used in English-style ales, where malt flavor is a dominant feature.

Adjuncts

Understanding Adjuncts

Adjuncts are unmalted grains or other fermentable materials added to the mash to supplement malted barley. They can lighten the body, add flavor, or reduce costs. Common adjuncts include corn, rice, barley, wheat, oats, and rye. Adjuncts must be cooked to gelatinize their starches before mashing, making them soluble and accessible to enzymatic breakdown.

Common Adjuncts and Their Impact

1. Corn: A popular adjunct in American brewing, corn adds sweetness and smoothness to beer. It lowers the protein and tannin content, reducing haze and lightening the body. Corn is often used in lagers to produce a clean, crisp profile. The sweetness of corn complements the hop bitterness in many light lagers, making them more palatable.
2. Rice: Used similarly to corn, rice imparts a light, clean flavor. It is a key ingredient in many mass-produced lagers, contributing to their mild taste and clarity. Rice reduces the protein and tannin content, lightening the body without adding significant flavor. This makes rice an excellent adjunct for creating beers with a smooth, clean finish.

3. Other Adjuncts:
   - Barley: Unmalted barley enhances malt character and can improve head retention.
   - Wheat: Improves head retention and mouthfeel. Wheat is a key ingredient in wheat beers and Belgian-style witbiers, also known as Belgian White. Witbiers are brewed with a significant portion of wheat in addition to barley malt. They are typically unfiltered, giving them a cloudy appearance. Often spiced with coriander and orange peel, witbiers have a refreshing, mildly spiced flavor profile with citrus notes.
   - Oats: Contribute creaminess and a smooth mouthfeel, commonly found in stouts. Oats add body and a velvety texture to the beer.
   - Rye: Adds a spicy note and can enhance the beer's body. Rye imparts a distinctive flavor that can add complexity to various beer styles.

Sugars

Role of Sugars in Brewing

Sugars are used in brewing to increase fermentable extract and adjust the flavor profile of the beer. While they are not typically used in lagers, they can enhance the complexity of ales. Different sugars contribute varying degrees of fermentability and flavor.

Types of Sugars

1. Corn Sugar (Dextrose): A highly fermentable sugar often used for priming beer—adding a small amount of sugar before bottling to carbonate the beer naturally. Dextrose ferments quickly and completely, adding alcoholic strength without residual sweetness. However, excessive use can lead to a thin body and a cidery off-flavor. Corn sugar is popular among homebrewers for its ease of use and predictability in fermentation.

2. Cane Sugar (Sucrose): Common table sugar used in brewing. Overuse can impart a cidery taste, but partially refined forms like light and dark brown sugar contribute a rum-like flavor that can be desirable in certain ale styles. The caramel and molasses notes from brown sugar add depth and complexity to darker ales.
3. Grain Sugars: Derived from the starches in grains like barley, corn, wheat, and rice. These sugars are produced through enzymatic conversion or acid hydrolysis and are used to increase alcohol content without significantly altering the flavor profile. Grain syrups and extracts provide fermentable sugars while imparting the grain's characteristic flavors.

Historical Context

Ancient Origins

The use of malt, adjuncts, and sugars in brewing dates back to ancient civilizations. The Sumerians and Egyptians utilized primitive malting techniques and local grains to produce fermented beverages. These early brewers relied on empirical observations, noting which methods produced desirable flavors and effects. The discovery of fermentation likely occurred by accident when wild yeasts fermented grain mixtures left out in the open. These ancient brewers observed the results and refined their techniques over time, creating the first rudimentary beers.

Medieval Europe

During the medieval period, monastic breweries in Europe refined brewing techniques. Monks meticulously recorded their processes in brewing logs, leading to improved consistency and quality. Brewing logs are detailed records that track ingredients, temperatures, times, and deviations during the brewing process, helping to identify patterns and improve techniques.

The use of different malts and adjuncts became more sophisticated, with brewers experimenting to achieve specific characteristics in their beers. Monasteries became centers of brewing knowledge, where the exchange of information contributed to the standardization of brewing practices. Monks also began to understand the importance of temperature control, maintaining consistent temperatures during mashing, fermentation, and storage to ensure optimal yeast performance and prevent off-flavors.

Modern Era

The 19th century brought scientific advancements that transformed brewing. The discovery of yeast's role in fermentation and the development of malting technology allowed for greater control over the brewing process. Modern brewers now have access to a wide variety of malts, adjuncts, and sugars, each with specific properties that can be manipulated to achieve desired outcomes.

The industrialization of brewing introduced precision and consistency, enabling large-scale production of high-quality beer. Advances in microbiology and chemistry have provided brewers with tools to analyze and optimize every aspect of the brewing process.

Rationalism—Deductive Reasoning

Applying Rationalism in Brewing

Rationalism, a philosophical approach that uses top-down deductive reasoning, plays a crucial role in understanding the biochemical principles behind malt, adjuncts, and sugars. By applying deductive reasoning, brewers can predict how different ingredients will affect the brewing process and the final product.

Biochemical Principles

1. Enzyme Activity: Understanding how enzymes like alpha-amylase break down complex starches into smaller dextrins, while beta-amylase converts these dextrins into fermentable sugars like maltose, allows brewers to make informed malt selections and optimize mashing conditions. Malts with high diastatic power are essential, especially when using adjuncts like rice or corn that lack their own enzymatic activity.

Deductive reasoning helps brewers achieve consistent starch conversion and control the fermentability of the wort.

2. Nitrogen and Protein Content: Recognizing that nitrogen content in malted barley directly correlates with its protein content is essential, as proteins are composed of nitrogen-containing amino acids. Using the formula Total Protein Content (%) = Nitrogen Content (%) × 6.25, brewers can ensure that their mash has adequate protein levels. This deductive approach helps avoid issues such as poor enzyme activity, excessive haze, and off-flavors caused by imbalanced protein content. Proper protein levels are crucial for optimal brewing performance and the quality of the final product.
3. Color and Flavor Development: Understanding the chemical changes during kilning allows brewers to select malts that impart desired characteristics to their beer. Kilning affects malt's color, flavor, and enzymatic activity. Higher temperatures produce darker malts with complex flavors but reduce enzyme activity, while lower temperatures preserve enzymatic activity and create lighter malts with milder flavors. Using deductive reasoning, brewers can choose malts that align with their brewing goals, ensuring the malt characteristics match the intended beer style and quality.

Empiricism—Inductive Reasoning

Applying Empiricism in Brewing

Empiricism involves gathering data through observation and experimentation. Brewers use inductive reasoning, a bottom-up logical process where generalizations are made based on observations and experimental data, to identify patterns and optimize brewing practices based on empirical evidence.

Refining Practices

1. Malt Modification: By observing the effects of different malting techniques on enzyme activity and protein levels, brewers can identify the best malts for their recipes. Empirical

data on how various malts perform under different conditions informs decisions about malt blends and mashing techniques. Continuous experimentation with malting processes leads to improvements in malt quality and performance.

For example, brewers might notice that a specific malting temperature consistently produces a desirable balance of enzyme activity and protein content. They then generalize this observation to refine their overall malting strategy, illustrating inductive reasoning.

2. Adjunct Use: Experimenting with different adjuncts allows brewers to understand their impact on flavor, body, and clarity. Empirical observations guide the use of adjuncts to achieve specific beer characteristics. For example, trials with rice and corn can reveal their effects on lightness and smoothness, while experiments with oats and rye can enhance body and flavor.

   By conducting these experiments and observing consistent results, brewers use inductive reasoning to generalize their findings. For instance, repeated trials showing that rice consistently lightens the beer body lead brewers to incorporate this adjunct into recipes where a lighter body is desired.

3. Sugar Additions: Testing different types and amounts of sugars helps brewers determine their effects on fermentation and flavor. Empirical data on sugar fermentability and impact on taste inform decisions about sugar additions. Controlled experiments with various sugars enable brewers to create balanced and complex ales without undesirable off-flavors.

   By observing the outcomes of these tests, brewers use inductive reasoning to generalize their findings. For example, if multiple trials show that using dark brown sugar adds a pleasant molasses flavor to ales, brewers may incorporate it into recipes where this characteristic is desired.

The Scientific Method—Six Elements

Advancing Brewing Techniques

The scientific method provides a structured approach to advancing brewing techniques. This systematic approach involves:

1. Observation: Monitor brewing processes and outcomes to identify areas for improvement and gather data for analysis.
2. Hypothesis Formation: Develop theories about potential improvements in brewing based on observations and existing knowledge.
3. Experimentation: Test different variables and conditions in brewing to discover if the results verify or falsify a predictive hypothesis.
4. Data Analysis with Assessment: Analyze the results to draw conclusions and refine brewing practices.
5. Replicability: Ensure that other brewers can obtain consistent results using the same methods and procedures in different breweries or laboratories, by different brewers.
6. Peer Review: The assessment of brewing techniques, experiments, and results by other brewers or experts to validate findings and improve practices.

Case Studies

Case Study 1: Malt

Optimizing Malt Modification

A brewery aims to improve the consistency of its pale ales by optimizing malt modification. They hypothesize that adjusting the germination time and kilning temperature could enhance the malt's enzymatic activity and flavor profile.

- Experimentation: Conduct controlled experiments with different germination periods and kilning temperatures.
- Data Collection: Measure enzyme levels, nitrogen content, and sensory characteristics.

- Data Analysis: Determine the optimal combination that produces malt with high diastatic power and desirable flavor attributes.
- Replicability and Peer Review: Share findings with other brewers, who replicate the results and validate the conclusions.
- Outcome: Establish a new standard for malt modification that improves the consistency and quality of the brewery's pale ales.

Case Study 2: Adjuncts

Incorporating Oats for Enhanced Mouthfeel

A craft brewery aims to create a smooth, creamy stout by incorporating oats into the mash. They hypothesize that using oats will enhance the beer's mouthfeel and body.

- Experimentation: Conduct trials with varying proportions of oats.
- Observation: Monitor the impact on viscosity, head retention, and flavor.
- Data Analysis: Identify the optimal percentage of oats that provides the desired creamy texture without overwhelming the malt profile.
- Replicability and Peer Review: Findings are replicated by others and assessed by experts.
- Outcome: Widespread adoption of oats in stout recipes, inspiring further experimentation with other adjuncts.

Case Study 3: Sugars

Balancing Cane Sugar Additions

A brewery specializing in Belgian-style ales seeks to refine its sugar addition practices to achieve a more balanced flavor profile. They hypothesize that using different types of cane sugar will affect the beer's fermentation dynamics and taste.

- Experimentation: Conduct trials with various forms of cane sugar, including light and dark brown sugar.
- Observation: Monitor fermentation rate, alcohol content, and sensory attributes.
- Data Analysis: Determine how each sugar type influences flavor and fermentation.
- Replicability and Peer Review: Other brewers replicate the experiments and validate the results.
- Outcome: Adjust sugar additions based on findings, achieving a well-balanced Belgian ale with complex flavors and smooth fermentation.

Challenges

Addressing Common Brewing Challenges

Brewers have faced numerous challenges throughout history, from maintaining consistency to preventing spoilage. Evidence-based solutions and troubleshooting tips have been developed to address these issues. Ongoing research and potential future directions continue to explore ways to improve brewing practices.

Haze Formation

- Problem: High protein and tannin levels can cause haze in beer.
- Solution: Use low-protein malts and adjuncts, along with proper mashing techniques.
- Techniques:
    - Employ Finings: Substances like isinglass, gelatin, or carrageenan (Irish moss) added to beer to clarify and stabilize the final product. Finings bind to unwanted particles like proteins, yeast, and polyphenols, causing them to precipitate and settle, improving clarity, taste, and stability.

- Cold Conditioning: The process of storing beer at low temperatures (32°F to 40°F or 0°C to 4°C) after fermentation. Cold conditioning encourages the precipitation of unwanted proteins and tannins, enhancing clarity and flavor by allowing flavors to mellow and mature.

Fermentation Problems

- Problem: Inconsistent yeast performance resulting from improper pitching rates or nutrient deficiencies.
- Solution: Optimize yeast propagation protocols and nutrient management.
- Techniques:
  - Adequate Oxygenation: Introduce oxygen into the wort before fermentation to help yeast build cell walls and reproduce. Essential during the initial yeast growth phase but should be avoided after fermentation starts to prevent oxidation.
  - Temperature Control: Maintain consistent fermentation temperatures to ensure optimal yeast performance and prevent off-flavors.

Flavor Imbalances

- Problem: Incorrect malt or adjunct selection can lead to undesirable flavors.
- Solution: Understand the properties of each ingredient and how they interact.
- Techniques:
  - Adjust Malt Blends: Based on empirical data, create malt combinations that achieve the desired flavor profiles.
  - Experiment with Adjuncts: Use inductive reasoning to refine adjunct use, ensuring they contribute positively to the beer's characteristics.

Conclusion

Malt, adjuncts, and sugars are fundamental components of brewing, each contributing unique properties that influence the final product. By understanding the historical context, biochemical principles, and practical applications of these ingredients, brewers can create beers with precise characteristics and consistent quality. The interplay between rationalism, empiricism, and the scientific method is crucial for advancing brewing practices and achieving optimal results.

The history of brewing is a rich tapestry that weaves together ancient traditions, medieval innovations, industrial advancements, and modern technological breakthroughs. From the earliest brews of Mesopotamia and Egypt to the refined techniques of medieval monks, and from the industrial-scale production of the 19th century to the craft beer revolution of today, the evolution of brewing reflects humanity's enduring passion for this remarkable beverage.

The continuous interplay of rationalism, empiricism, and the scientific method has driven the development of brewing practices, ensuring that beer remains a dynamic and exciting field. As we look to the future, the integration of tradition and innovation promises to keep the world of beer vibrant and evolving.

> *To brew with malt, adjuncts, and sugars is not merely to manipulate ingredients—it is to take moral responsibility for transformation itself. Every decision, from the choice of grain to the degree of modification, reflects a philosophy of care: that creation ought to be deliberate, informed, and respectful of its origins. In honoring tradition with critical thought and empirical observation, the brewer becomes an ethical agent—translating knowledge into nourishment, and process into purpose.*

**

Review Questions

True/False Questions

1. True or False: The heart of brewing lies in the fundamental ingredients of malt, adjuncts, and sugars, which are the building blocks of beer.

2. True or False: Malt is primarily derived from wheat and undergoes germination and kilning to become malt.

3. True or False: Diastatic power is a measurement of the malt's enzyme content and is crucial for efficient starch conversion.

4. True or False: Adjuncts are unmalted grains or other fermentable materials added to the mash to supplement malted barley.

5. True or False: Corn sugar is often used for priming because it ferments quickly and completely, adding alcoholic strength without leaving residual sweetness.

Multiple Choice Questions

6. Which process in malt production involves soaking barley in water to trigger germination?
A) Kilning
B) Mashing
C) Steeping
D) Boiling

7. Which of the following is a common adjunct used in American brewing to add sweetness and smoothness to beer?
A) Rye
B) Wheat
C) Corn
D) Oats

Brewer Vignettes

8. Brewer Vignette 1: You are a brewer in a modern brewery looking to optimize your pale ale production. You hypothesize that adjusting the germination time and kilning temperature could enhance the malt's enzymatic activity and flavor profile. Describe the process you would follow and the potential results.
A) Ignore germination time and focus only on kilning temperature.
B) Conduct controlled experiments with different germination periods and kilning temperatures, then analyze enzyme levels, nitrogen content, and sensory characteristics.
C) Use the same germination time and kilning temperature for all batches.
D) Only vary the germination time without changing the kilning temperature.

9. Brewer Vignette 2: As a craft brewer, you want to create a smooth, creamy stout by incorporating oats into the mash. Explain how you would experiment with varying proportions of oats and what characteristics you would monitor.
A) Only use oats without considering proportions.
B) Experiment with varying proportions of oats, monitoring viscosity, head retention, and flavor.
C) Use oats at a fixed proportion in all batches.
D) Focus on other ingredients and ignore oats.

10. Brewer Vignette 3: You aim to refine your sugar addition practices to achieve a balanced flavor profile in your Belgian-style ale. Describe how you would test different types and amounts of sugars and what you would look for in the results.
A) Use only one type of sugar in all batches.
B) Conduct trials with various forms of cane sugar, monitoring fermentation rate, alcohol content, and sensory attributes.
C) Avoid using sugars in the brewing process.
D) Use random amounts of sugar without testing.

**

Correct Answers

True/False Questions

1. True
2. False (Malt is primarily derived from barley.)
3. True
4. True
5. True

Multiple Choice Questions

6. C) Steeping
7. C) Corn

Brewer Vignettes

8. B) Conduct controlled experiments with different germination periods and kilning temperatures, then analyze enzyme levels, nitrogen content, and sensory characteristics.
9. B) Experiment with varying proportions of oats, monitoring viscosity, head retention, and flavor.
10. B) Conduct trials with various forms of cane sugar, monitoring fermentation rate, alcohol content, and sensory attributes.

**

Beyond The Chapter

- Bamforth, Charles W. "Beer: Tap into the Art and Science of Brewing." Oxford University Press, 2009.
  - *Provides insights into the role of malt and adjuncts in brewing and their impact on beer quality.*
- Fix, George. "Principles of Brewing Science: A Study of Serious Brewing Issues." Brewers Publications, 1999.
  - *Explains the scientific principles behind the use of malt, adjuncts, and sugars in brewing.*
- Palmer, John J. "How to Brew: Everything You Need to Know to Brew Great Beer Every Time." Brewers Publications, 2017.

    - *Covers practical aspects of using malt and sugars in brewing and their effects on beer flavor.*
- White, Chris, and Jamil Zainasheff. "Yeast: The Practical Guide to Beer Fermentation." Brewers Publications, 2010.
    - *Discusses how different sugars and adjuncts influence yeast performance and fermentation outcomes.*
- Brewers Association and American Homebrewers Association websites for articles, forums, and educational resources.
    - *Provides additional resources and discussions on the use of malt, adjuncts, and sugars in brewing.*

Weblinks

- Malt in Brewing - Brew Your Own: https //byo.com/?s=Malt&post_type=article&orderby=relevance
- Understanding Adjuncts - Craft Beer & Brewing Magazine: https://beerandbrewing.com/search/?search=Adjuncts
- The Role of Sugars in Brewing - Brewers Association: https://www.brewersassociation.org/?s=sugar
- Adjuncts in Brewing - Homebrew Academy: https://homebrewacademy.com/?s=Adjuncts
- Historical Use of Adjuncts - American Homebrewers Association: https://www.homebrewersassociation.org/?s=Adjuncts
- Adjunct Grains - Northern Brewer: https://www.northernbrewer.com/pages/search-results-page?q=adjunct%20grains

These references offer comprehensive insights into Malt, Adjuncts, and Sugars serving as valuable resources for both homebrewers and professional brewers.

***

Chapter 3: The Silent Key—Mastering Water Chemistry in Brewing

In the art of brewing, water is the silent key, unlocking the true potential of every ingredient.

Imagine standing beside a tranquil stream, the clear water reflecting the sky above. This serene image belies the critical importance of water in brewing. Often referred to as the "forgotten ingredient," water is the foundation upon which all great beers are built. Its chemical composition—known as water chemistry—can significantly impact the flavor, mouthfeel, and stability of the beer.

Water Chemistry is the study of the chemical composition of water, including pH levels and mineral content, and its effects on brewing. To truly master the craft of brewing, one must understand the complexities of water chemistry, including pH levels, mineral content, and various water treatment techniques. Proper pH levels and mineral balances are crucial in various stages of brewing, including mashing, fermentation, and finishing. Monitoring and adjusting these factors is essential for producing high-quality beer.

This chapter will delve into these key concepts, connecting them to the broader themes of rationalism, empiricism, and the scientific method. By examining the historical evolution of water treatment in brewing, we will see how theoretical frameworks and empirical observations have shaped modern practices.

Ancient Origins

The significance of water in brewing has been recognized since ancient times. Early brewers understood that the quality of their local water source had a profound effect on the beer they produced. In ancient Mesopotamia and Egypt, brewers relied on empirical observations to select the best water for brewing. They knew, for example, that water from certain wells or rivers yielded better-tasting beer, even if they didn't fully understand the chemistry behind it. These early brewers' empirical methods laid the foundation for understanding the importance of water quality in brewing.

Medieval Europe

As brewing spread to Europe, the influence of local water became even more pronounced. In the medieval period, monastic brewers meticulously documented their methods, including their water sources. The soft water of Pilsen in the Czech Republic became legendary for its suitability in brewing pale lagers. Soft water is characterized by low mineral content, often preferred for brewing certain beer styles like lagers, as it allows delicate malt and hop flavors to shine without interference from mineral flavors.

Conversely, the hard water of Burton-on-Trent in England was perfect for brewing hoppy ales. Hard water has high mineral content, particularly calcium and magnesium, which can affect beer flavor and brewing processes. Burton-on-Trent is a town in England known for its high sulfate water, ideal for brewing pale ales, and influential in the development of water treatment techniques. The high sulfate content enhances the perceived dryness and bitterness of beer, making it beneficial for hoppy styles.

These observations led to an empirical understanding that water chemistry must be tailored to the beer style being brewed. This historical knowledge highlights the critical role that water has played in shaping regional brewing traditions.

Industrial Revolution

The Industrial Revolution brought significant advancements in water treatment. Brewers began to understand the importance of water chemistry more scientifically. One milestone was the recognition of the role of minerals such as calcium and sulfate.

Calcium is a mineral in brewing water that helps lower wort pH, assists enzyme activity, and supports yeast health. Ideal levels are between 50-100 ppm (parts per million). Calcium also plays a crucial role in protein coagulation during the boil, improving beer clarity and stability.

Sulfate (SO_4^{2-}) enhances the perceived dryness and bitterness of beer, crucial for hoppy beer styles like IPAs, where levels above 200 ppm are common. It is added through salts such as calcium sulfate (gypsum) and magnesium sulfate (Epsom salt). Proper sulfate levels help achieve the desired flavor balance and hop character, while excessive amounts can cause harshness.

Brewers in Burton-on-Trent learned to replicate their water's unique mineral profile by adding gypsum (calcium sulfate) to increase calcium and sulfate levels, enhancing hop bitterness and dryness. This rational approach to water treatment marked a shift from empirical observations to systematic scientific methods. Brewers began using salts—various mineral compounds dissolved in water—to adjust its chemical profile. Common brewing salts include calcium sulfate (gypsum) to lower pH and enhance hop bitterness, and calcium chloride to lower pH and enhance sweetness.

The advancements made during this period set the stage for modern brewing techniques, demonstrating how theoretical understanding could be applied to improve brewing outcomes consistently.

Modern Era

In the modern era, the scientific understanding of water chemistry has advanced significantly. The craft beer movement, with its emphasis on quality and innovation, has driven further exploration into water treatment techniques. Brewers now routinely test and adjust their water profiles to match the specific requirements of different beer styles.

This ongoing evolution reflects a blend of rationalism, empiricism, and the scientific method, ensuring that water quality is optimized for every brew. Modern brewers have the tools and knowledge to fine-tune their water to achieve desired results, reflecting the deep integration of science and craft in brewing.

Rationalism—Deductive Reasoning

Rationalism has played a crucial role in the development of water treatment techniques. Theoretical frameworks and established principles have guided the understanding and application of water chemistry in brewing. This approach has allowed brewers to apply consistent, reliable methods to improve their brewing process.

Theoretical Frameworks

One key aspect of rationalism in brewing water treatment is the application of water chemistry principles. For example, the understanding that calcium helps lower wort pH and assists enzyme activity is based on deductive reasoning. Brewers apply this knowledge by adding calcium salts to their water to achieve the desired pH levels and enzyme function.

Buffers are chemical substances that stabilize the pH level of the mash and wort by neutralizing acids and bases. Common buffers include sodium bicarbonate, which raises pH if it gets too low, and phosphates, which help maintain a stable pH. To lower pH, brewers often use food-grade acids like lactic acid or phosphoric acid. Proper buffering is crucial for optimal enzyme activity during mashing, yeast health during fermentation, and overall beer quality.

Historical Example

The Burton-on-Trent brewers' use of gypsum to replicate their local water's mineral content is a prime example of rationalism. By understanding the theoretical benefits of calcium sulfate, they could improve the quality of their beer consistently. This top-down approach, guided by established principles, has been integral to the evolution of water treatment in brewing.

Empiricism—Inductive Reasoning

Empiricism has been equally important in refining water treatment practices. Brewers have relied on observations and experiments to enhance their understanding of water chemistry. This bottom-up approach allows for practical improvements based on real-world results, leading to a more nuanced understanding of brewing processes.

Observational Learning

Throughout history, brewers have made empirical observations about their water sources. For instance, the discovery that the soft water of Pilsen produced superior lagers compared to harder water sources was based on careful observation and comparison. These empirical findings have been critical in developing water treatment practices tailored to specific brewing needs.

Experimental Results

Empirical data has driven many advancements in water treatment. Brewers have conducted experiments to determine the effects of various minerals on beer flavor and stability. For example, the role of chloride in enhancing malt sweetness and fullness was identified through empirical testing and tasting. Chloride levels between 50-150 ppm are ideal for enhancing malt character without causing off-flavors.

The Scientific Method—Six Elements

The scientific method has been instrumental in advancing water treatment techniques in brewing. This approach involves systematic observation, hypothesis formation, experimentation for verifying or falsifying the hypothesis, data analysis with assessment, replicability, and peer review. Applying the scientific method ensures that brewing practices are continually improved based on rigorous testing and analysis.

Case Study: Dechlorination Methods

A notable example is the work done on dechlorination methods. Chlorine and chloramine (a combination of chlorine and ammonia) are disinfectants added to municipal water supplies, which can cause off-flavors in beer if not removed. Brewers observed that chlorine in water caused medicinal or plastic-like off-flavors, leading to hypotheses about how to remove it effectively.

Through experimentation, methods such as activated carbon filters and Campden tablets (potassium metabisulfite) were developed and tested. Activated carbon filters adsorb contaminants onto the porous surface of activated carbon, effectively removing chlorine and chloramine. Campden tablets neutralize chlorine and chloramine through a chemical reaction. Data analysis confirmed their efficacy, and these findings were shared and peer-reviewed within the brewing community, leading to widespread adoption of these techniques.

This collaborative approach exemplifies the power of the scientific method in solving practical brewing challenges.

Water Chemistry in Detail

pH Levels

The pH (Potential of Hydrogen) of brewing water affects enzyme activity during mashing, with the optimal range being between 5.2 and 5.6. pH is a measure of the acidity or alkalinity of a solution, on a scale from 0 to 14. Proper pH levels ensure optimal enzyme activity during mashing, promote yeast health and efficient fermentation, and enhance the stability and flavor of the final beer.

Adjusting water pH using food-grade acids or bases ensures enzymes function efficiently, leading to better starch conversion and higher-quality beer. Maintaining the correct pH is essential for achieving desired flavors and efficiencies in brewing.

Mineral Content

The mineral composition of water influences various aspects of brewing and the final beer. Key minerals include:

- Calcium: Helps lower wort pH, assists enzyme activity, and aids in yeast health. Ideal levels are between 50-100 ppm. Calcium also plays a crucial role in protein coagulation during the boil, improving beer clarity and stability. During the boiling phase, heat causes water molecules to detach from larger proteins in the wort—a process known as denaturing. This makes the large protein molecules insoluble, leading them to flocculate (clump together) in a process called the hot break. Achieving a hot break is challenging if the wort's pH is lower than 5.0.
- Magnesium: Essential for yeast health but can impart a bitter flavor at high levels. The optimal range is 10-30 ppm.
- Sodium: Adds smoothness and enhances mouthfeel, with optimal levels between 50-150 ppm.
- Potassium: Can inhibit certain enzymes in the mash and should be used cautiously in brewing water.
- Chloride: Enhances malt sweetness and fullness, with ideal levels between 50-150 ppm.
- Sulfate: Enhances hop bitterness and dryness, crucial for hoppy beer styles like IPAs. Levels above 200 ppm are common in these styles but should be balanced to avoid harshness.

Brewing Salts and Buffers

Brewing salts are minerals added to water to adjust its chemical composition, enhancing the brewing process and the final beer's flavor profile. Common brewing salts include:

- Calcium Sulfate (Gypsum): Used to increase calcium and sulfate levels, enhancing hop bitterness and dryness. It can be added to the mash to lower pH, with a maximum of 1 teaspoon per 5 gallons.

- Calcium Chloride: Adds calcium and chloride ions, enhancing malt sweetness and fullness. It can also lower pH when added to the mash.
- Magnesium Sulfate (Epsom Salts): Adds magnesium and sulfate, influencing beer flavor and mouthfeel. It serves as a yeast nutrient but should be used sparingly to avoid bitterness.
- Sodium Bicarbonate (Baking Soda): Raises pH if it gets too low, but excessive use can lead to soapy flavors.

Buffers help control pH levels, promoting yeast health and optimizing enzyme activity during mashing. Proper use of brewing salts and buffers ensures that the water chemistry matches the desired beer style, optimizing overall beer quality.

Water Treatment Techniques

Various techniques are employed to treat brewing water, ensuring it is suited for the desired beer style. These techniques combine scientific principles and practical know-how to optimize brewing conditions.

Dechlorination

Removing chlorine and chloramine is essential to avoid off-flavors and protect yeast health. Methods include:

- Activated Carbon Filters: These filters adsorb contaminants onto the activated carbon's porous surface, effectively removing chlorine, chloramine, and impurities, enhancing beer taste and quality.
- Campden Tablets (Potassium Metabisulfite): Tablets used to neutralize chlorine and chloramine in brewing water through a chemical reaction.

Adjusting Mineral Content

Understanding the water profile of classic beer regions helps brewers replicate traditional styles more accurately. The specific composition of minerals and pH in a water source influences the brewing process and the final beer.

Brewers adjust mineral content using brewing salts:

- Calcium Sulfate (Gypsum): Enhances hop bitterness and dryness, ideal for pale ales and IPAs.
- Calcium Chloride: Enhances malt sweetness and fullness, suitable for malty beer styles.
- Epsom Salts (Magnesium Sulfate): Adds magnesium and sulfate, influencing yeast health and hop character.

By adjusting the mineral content, brewers can enhance specific characteristics of their beer, such as flavor, mouthfeel, and aroma.

Challenges in Water Treatment

Water Testing

Regularly testing your water source for pH and mineral content using a water test kit or sending samples to a lab for detailed analysis is crucial for making informed adjustments to your brewing water. Accurate testing ensures that brewers can maintain consistency and quality in their beer.

Consistency

Maintaining consistent water quality by using the same source and treatment methods for each batch helps ensure that each batch of beer meets the same high standards, reducing variability in the final product. Consistency in water treatment is essential for producing reliable and predictable results in brewing.

Common Brewing Issues

Brewers face numerous challenges related to the use of water in brewing. Issues such as inconsistent fermentation, off-flavors, and spoilage are common. Addressing these problems requires careful observation and experimentation.

Evidence-Based Solutions

Effective solutions, such as maintaining proper sanitation practices and controlling fermentation temperatures, are informed by empirical evidence and scientific principles. Ongoing research continues to explore new ways to address these challenges and improve brewing practices. This continuous cycle of problem-solving and innovation drives the evolution of brewing techniques.

Conclusion

The chemistry and treatment of brewing water are complex but essential aspects of the brewing process. By understanding and applying rational principles, empirical observations, and the scientific method, brewers can optimize their water to produce high-quality beer consistently.

This chapter underscores the importance of water in brewing and sets the stage for exploring other critical brewing ingredients in subsequent chapters. Mastering water chemistry is a fundamental step in becoming a proficient brewer and ensuring the highest quality in every brew.

> *To master water in brewing is to practice an ethic of attention—to that which is silent, foundational, and often overlooked. It is a reminder that excellence is built not only on what is visible, but on what is carefully balanced beneath. The brewer's responsibility lies not in domination, but in stewardship: to know, to respect, and to align with nature's quiet structures. In tending even the subtlest variables, we act not just as technicians, but as moral agents, honoring the unseen depths that make excellence possible.*

**

Review Questions

True/False Questions

1. True or False: Water's chemical composition significantly impacts the flavor, mouthfeel, and stability of beer.

2. True or False: The soft water of Burton-on-Trent in England is perfect for brewing pale lagers.

3. True or False: Rationalism with its top down deductive reasoning have played a crucial role in the development of water treatment techniques in brewing.

4. True or False: Empiricism involves gathering data through observation and experimentation, which has been essential in refining water treatment practices in brewing.

5. True or False: The pH of brewing water should ideally be between 5.2 and 5.6 to ensure optimal enzyme activity during mashing.

Multiple Choice Questions

6. Which mineral is essential for yeast health but can impart a bitter flavor at high levels?
A) Calcium
B) Sodium
C) Magnesium
D) Chloride

7. Which technique is commonly used to remove chlorine from brewing water?
A) Freezing
B) Boiling
C) Filtration through activated carbon
D) Fermentation

Brewer Vignettes

8. Brewer Vignette 1: Imagine you are a brewer trying to optimize the pH level of your brewing water. You hypothesize that adding calcium salts will help achieve the desired pH range. Describe the steps you would take to test this hypothesis and the expected results.
A) Use random amounts of calcium salts without testing pH levels.
B) Ignore pH levels and focus on temperature.
C) Avoid using any additives and rely solely on the water source's natural pH.
D) Add calcium salts, monitor pH levels, and adjust quantities based on the results to achieve a pH between 5.2 and 5.6.

9. Brewer Vignette 2: As a brewer, you want to replicate the mineral profile of Burton-on-Trent's water for your pale ale. Explain how you would adjust your water chemistry and what effects you expect.
A) Increase the sodium content significantly.
B) Use soft water with minimal mineral content.
C) Add gypsum (calcium sulfate) to enhance hop bitterness and dryness, reflecting Burton-on-Trent's water profile.
D) Focus on increasing chloride levels for malt sweetness.

10. Brewer Vignette 3: You notice off-flavors in your beer and suspect chlorine in the water supply is the cause. Describe the methods you would use to remove chlorine and how you would test their effectiveness.
A) Boil the water for an extended period.
B) Add more hops to mask the off-flavors.
C) Ignore the issue and continue brewing with the same water.
D) Use activated carbon filters or Campden tablets (potassium metabisulfite) to neutralize chlorine and test the water before and after treatment.

**

Correct Answers

True/False Questions

1. True
2. False (The soft water of Pilsen in the Czech Republic is perfect for brewing pale lagers; Burton-on-Trent has hard water suitable for hoppy ales.)
3. True
4. True
5. True

Multiple Choice Questions

6. C) Magnesium
7. C) Filtration through activated carbon

Brewer Vignettes

8. D) Add calcium salts, monitor pH levels, and adjust quantities based on the results to achieve a pH between 5.2 and 5.6.
9. C) Add gypsum (calcium sulfate) to enhance hop bitterness and dryness, reflecting Burton-on-Trent's water profile.
10. D) Use activated carbon filters or Campden tablets (potassium metabisulfite) to neutralize chlorine and test the water before and after treatment.

**

Beyond The Chapter

- Bamforth, C. W. (2003). Beer: Tap into the Art and Science of Brewing. Oxford University Press.
    - *Explores the importance of water quality and treatment in brewing.*
- Palmer, J. (2006). How to Brew: Everything You Need To Know To Brew Beer Right The First Time. Brewers Publications.
    - *Provides practical guidance on water treatment and its impact on brewing.*
- Kaminski, M., & Hieronymus, S. (2013). Water: A Comprehensive Guide for Brewers. Brewers Publications.

    - *A detailed guide on water chemistry and treatment methods specific to brewing.*
- DeLange, A. (2017). The Practical Brewer: A Manual for the Brewing Industry. Master Brewers Association of the Americas.
    - *Offers practical advice on managing water quality and treatment in commercial brewing.*
- Daniels, R. (1996). Designing Great Beers: The Ultimate Guide to Brewing Classic Beer Styles. Brewers Publications.
    - *Discusses the role of water in shaping the characteristics of different beer styles.*

Weblinks

- Water Chemistry for Brewers - Craft Beer & Brewing Magazine: https://beerandbrewing.com/search/?search=Water
- Understanding Brewing Water - Brew Your Own: https://byo.com/?s=water&post_type=any&orderby=relevance
- Water Treatment Basics - American Homebrewers Association: https://www.homebrewersassociation.org/?s=water
- The Science of Brewing Water - MoreBeer: https://www.morebeer.com/search?search=Water
- The Importance of Water Quality - Brewers Association: https://www.brewersassociation.org/?s=water
- Water and its Effects on Beer - Homebrew Academy: https://homebrewacademy.com/?s=Water

These references provide deeper insights into the chemistry and treatment of brewing water, offering valuable resources for both homebrewers and professionals.

***

Chapter 4: The Crucible of Mashing—Where Grains and Enzymes Converge

Mashing is the crucible where grains and enzymes converge, setting the stage for every beer's unique character.

Imagine the comforting aroma of grains as they steep in hot water, slowly transforming into a fragrant, sugary liquid. This is the mashing process, a pivotal step in brewing that sets the stage for creating beer. Mashing is where the alchemy happens, where raw ingredients are converted into a fermentable solution that yeast will later turn into alcohol and carbonation.

In brewing, alcohol refers to ethanol, produced by yeast during fermentation as they convert sugars into alcohol and carbon dioxide. It is measured as a percentage by volume (ABV) and influences the beer's flavor, body, and sensory experience. The fermentation process and the amount of fermentable sugars determine the final alcohol level.

Carbonation is the process of dissolving carbon dioxide (CO_2) into beer, creating bubbles and giving the beer its fizzy texture. Carbonation can be achieved naturally through priming—adding a small amount of sugar before bottling to allow yeast to produce CO_2—or artificially through force carbonation using pressurized CO_2 tanks.

This transformative process begins with mashing-in, the initial step of mashing, where crushed malt is combined with hot water to create a mash. The mash water should start at a precise temperature and a pH around 7.0, which will result in a mash pH between 5.2 and 5.6, ensuring optimal enzyme activity. With the proper temperature and pH, enzymes in the malt become active, breaking down complex starches into simpler, fermentable sugars—a process known as saccharification. This biochemical transformation is foundational to brewing, determining much of the beer's final character.

The Precision of Mashing-In

Mashing-in is not merely a mechanical process but a precise and deliberate action that lays the groundwork for the entire brewing journey. The crushed malt, often referred to as grist, is carefully mixed with hot water, typically at a temperature between 148°F and 158°F (64°C and 70°C). The exact temperature depends on the desired beer style, as different temperatures activate different enzymes.

To achieve this, brewers calculate the strike temperature—the temperature of the water added to the crushed malt during mashing-in, usually higher than the desired mash temperature to account for heat loss from the cooler malt. For example, if aiming for a mash temperature of 152°F (67°C), the strike temperature might be around 163°F (73°C).

A higher mashing temperature tends to create a fuller-bodied beer with more unfermentable sugars, resulting in a thicker, maltier beer. Conversely, a lower temperature produces a lighter, drier beer with more fermentable sugars, leading to higher alcohol content.

To monitor and control these temperatures accurately, brewers use a thermometer—an essential instrument used to measure the temperature of the mash, boiling wort, and fermentation. Precise temperature control is crucial for enzyme activity and, ultimately, the quality of the beer.

The quality of the mash hinges on maintaining optimal conditions for enzyme activity. Enzymes are highly sensitive to both temperature and pH, so brewers must ensure these variables are kept within ideal ranges. The typical mash pH is between 5.2 and 5.6—the acidity or alkalinity of the mash is critical for optimal enzyme activity and flavor stability. Deviations from this range can inhibit enzyme activity, leading to poor conversion of starches to sugars. The delicate balance of these factors requires a deep understanding of the biochemical processes at play, as well as careful attention to detail during the mashing-in process.

Historical Foundations of Mashing

Ancient Origins

The practice of mashing dates back thousands of years, with early evidence found in ancient civilizations such as Mesopotamia and Egypt. These early brewers discovered through empirical observation that steeping grains in water produced a sweet liquid that could be fermented into beer. Although they lacked a scientific understanding of enzymes, their empirical trial-and-error approach laid the groundwork for modern mashing techniques. They rationally understood that specific conditions produced better results, even if the reasons and specifics of why and how were not entirely clear.

Medieval Europe

As brewing knowledge spread across Europe, different regions developed their own mashing practices based on local ingredients and brewing traditions. In medieval Europe, monastic brewers played a crucial role in refining these techniques. They meticulously recorded their methods, contributing significantly to the body of brewing knowledge. The gradual accumulation of empirical data and practical experience allowed brewers to improve their mashing processes over time. This era laid the foundation for the structured approaches seen in modern brewing.

Industrial Revolution

The Industrial Revolution brought about significant advancements in mashing techniques. The development of the thermometer and hydrometer enabled brewers to monitor temperature and specific gravity with greater precision. The specific gravity is a measure of the density of beer relative to water, used to determine the amount of fermentable sugars and the beer's alcohol content. It is measured with a hydrometer or refractometer.

This period also saw the introduction of mechanical stirring devices, equipment used to maintain even temperature distribution and prevent the grain bed from compacting during lautering. Lautering is the process of separating the liquid wort from the solid grain husks after mashing, critical for ensuring beer clarity, flavor, and quality. These innovations were driven by both rationalism and empiricism, as brewers applied scientific principles to enhance their practices. The era marked a shift from traditional, experience-based methods to more scientifically informed techniques.

Rationalism—Deductive Reasoning in Mashing

Rationalism has played a vital role in understanding the biochemical principles underlying mashing. Theoretical frameworks and established principles have guided the development of mashing techniques through deductive reasoning. This approach allows brewers to apply general principles to specific scenarios, ensuring consistency and efficiency.

Understanding Enzyme Activity

The study of enzymes and their optimal conditions for activity is a prime example of rationalism in brewing. Alpha-amylase and beta-amylase are two key enzymes in the mashing process.

- Alpha-amylase is active at temperatures up to 159°F (158-162°F or 70-72°C) and a pH of around 5.7. This enzyme breaks down complex starches into a mixture of fermentable sugars and dextrins, creating a less fermentable wort and, consequently, a thicker, maltier beer. Alpha-amylase works quickly to reduce long starch chains into shorter pieces, contributing to the beer's body and mouthfeel.
- Beta-amylase is most active at temperatures around 149°F (140-149°F or 60-65°C) and a pH of 5.3-5.4. It breaks down starches into maltose, a highly fermentable sugar, resulting in a more fermentable wort. Beta-amylase is deactivated more rapidly at higher temperatures, so careful temperature control

is essential. This enzyme produces a wort that leads to beers with higher alcohol content and a drier finish.

By understanding these enzymes' temperature and pH requirements, brewers can design mashing schedules that optimize their activity, resulting in a more efficient conversion of starches to sugars. This theoretical knowledge allows brewers to predict outcomes and adjust processes accordingly.

Theoretical Frameworks in Mashing

The development of step mashing, where the temperature of the mash is gradually increased to activate different enzymes at their optimal temperatures, is another example of rationalism in brewing. This technique allows brewers to control the beer's body and fermentability more precisely. The application of these principles demonstrates how deductive reasoning has shaped modern mashing practices. Rationalism has provided a solid foundation for developing techniques that enhance the quality and consistency of beer.

Empiricism—Inductive Reasoning in Mashing

Empiricism has been equally important in refining mashing practices. Brewers have relied on observations and experiments to enhance their understanding of mashing. This approach allows for the continuous improvement of techniques based on practical experience and evidence.

Observational Learning

Throughout history, brewers have made empirical observations about the mashing process. For example, they noted that maintaining certain temperatures resulted in better sugar conversion and more consistent beer quality. These observations led to the development of mashing techniques that optimized enzyme activity. The practice of carefully observing and recording results has been a cornerstone of brewing advancement.

Experimental Results

Empirical data has driven many advancements in mashing. Brewers have conducted experiments to determine the effects of different mashing temperatures and pH levels on enzyme activity and sugar conversion. For instance, they discovered that a lower mashing temperature favored beta-amylase activity, producing a highly fermentable wort, while a higher temperature favored alpha-amylase, resulting in a fuller-bodied beer.

This practice of continuous improvement—the ongoing practice of refining brewing techniques and processes to achieve consistently high-quality results, often through meticulous record-keeping, experimentation, and learning—demonstrates the importance of inductive reasoning in brewing. Experimentation continues to refine and enhance mashing techniques, ensuring that brewing remains both an art and a science.

The Scientific Method—Advancing Mashing Techniques

The scientific method has been instrumental in advancing mashing techniques in brewing. This approach involves systematic observation, hypothesis formation, experimentation for verifying or falsifying the hypothesis, data analysis with assessment, replicability, and peer review. Applying the scientific method ensures that brewing practices are continually improved based on rigorous testing and analysis.

Case Study: Single Infusion Mashing

A notable example is the development of single infusion mashing. Brewers observed that maintaining a constant temperature throughout the mashing process produced consistent results. They hypothesized that this method would be effective for well-modified malts, which contain sufficient enzyme activity to convert starches at a single temperature.

Experiments confirmed this hypothesis, and the data was analyzed and shared within the brewing community, leading to widespread adoption of single infusion mashing. This process highlights the importance of scientific inquiry in brewing.

The Mashing Process in Detail

Mashing In

Mashing-in involves combining crushed malt with hot water to create a mash. This step requires precision and attention to detail, as the temperature and pH of the mash must be within the optimal ranges for enzyme activity. The typical temperature range for mashing-in is between 148°F and 158°F (64°C and 70°C), while the pH should be between 5.2 and 5.6.

The Mash Tun is the vessel used for mixing crushed malt with hot water during the mashing process. It is equipped to maintain temperature and sometimes facilitate heating, providing an environment where the conversion of starches to sugars occurs.

Achieving these conditions is crucial for effective starch conversion. The use of precise thermometers and careful calculation of the strike temperature ensures that the mash reaches and maintains the desired temperature.

Enzyme Activity

The roles of alpha-amylase and beta-amylase are critical in the mashing process. The interaction between these enzymes, moderated by temperature and pH, determines the balance of fermentable and unfermentable sugars in the wort—the liquid extracted from the mashing process. This balance influences the beer's final sweetness, body, and alcohol content. Understanding this balance is key to crafting beer with desired characteristics.

Advanced Mashing Techniques

Brewers have developed various mashing techniques to control the beer's characteristics more precisely. These methods allow for greater flexibility and creativity in brewing.

Single Infusion Mash

Single infusion mash maintains a constant temperature throughout the mashing process. It is straightforward and suitable for well-modified malts. This method is prevalent among homebrewers and is ideal for producing a wide range of beer styles with consistent quality. It simplifies the mashing process while ensuring effective enzyme activity.

Single Infusion Mash (Kettle or Cooler)

1. Preheat the Insulated Mash Tun: Preheat the insulated Mash Tun with 180°F (82°C) water for 20 minutes, then return the water to the kettle or hot liquor tank. Mash water should have a pH around 7.0 but no lower. If necessary, add calcium to lower pH.
2. Calculate and Adjust the Water-to-Grain Ratio: The water-to-grain ratio affects the mash's consistency and efficiency. Commonly, brewers use around 1.25 to 1.5 quarts of water per pound of grain. This ratio influences the thickness of the mash and the effectiveness of starch conversion.
3. Fill Mash Tun with Strike Water: Fill the Mash Tun with water at a strike temperature 11°F (6°C) higher than desired mash temperature, for example, 165°F (74°C).
4. Mash-In Grist: Add the crushed malt (grist) to the water, correcting the temperature to the desired mash temperature of 150-158°F (66-70°C) by tempering with hot or cold water.
5. Correct Mash pH: Adjust the mash pH to between 5.2 and 5.6. To raise pH, add calcium carbonate ½ teaspoon at a time, up to a maximum of 2 teaspoons per 5 gallons. Calcium carbonate increases alkalinity when the mash is too acidic. To lower pH, add calcium chloride or calcium sulfate (gypsum) ½ teaspoon at a time, up to a maximum of 1 teaspoon per 5 gallons.
6. Hold Temperature: Maintain the mash temperature for at least 1 to 1½ hours, stirring every half hour. Use precise thermometers and regularly calibrate equipment to ensure

temperature accuracy. Keep the Mash Tun covered except when stirring.

7. Mash-Out: Mash-out by boosting the temperature to 165-168°F (74-76°C) if using a Mash Tun or Lauter Tun that can be heated. Hold for 5 minutes. This step halts enzyme activity, stabilizes the sugar profile, and makes the mash less viscous, facilitating easier lautering and extraction of wort.

Step Mashing

Step mashing involves gradually increasing the temperature to activate different enzymes at their optimal temperatures. Starting at a lower temperature, the mash is heated in steps, allowing various enzymes to work on the starches sequentially. This method provides greater control over the beer's body and fermentability, making it useful for producing complex beers with precise characteristics.

Step Mashing Overview

Step mashing involves multiple temperature rests during the mashing process, each activating different enzymes that break down specific components in the malt.

Temperature Rests:

1. Acid Rest (95-113°F or 35-45°C): An acid rest activates phytase enzymes, which break down phytin in the malt into phytic acid and minerals. This reduces the pH of the mash, improving enzyme activity, protein degradation, and overall beer clarity and stability.
2. Protein Rest (122°F or 50°C): During the protein rest, proteolytic enzymes break down proteins into amino acids and peptides. This process improves head retention and reduces haze in the finished beer. An over-long protein rest can cause problems with foam retention and a thin body, so it's recommended to keep it to half an hour or less.

3. Beta-Amylase Rest (140-149°F or 60-65°C): Beta-amylase is most active in this temperature range, breaking down starches into maltose, a highly fermentable sugar. This rest produces a more fermentable wort, leading to a drier beer with higher alcohol content.
4. Alpha-Amylase Rest (158-162°F or 70-72°C): Alpha-amylase works at higher temperatures, breaking down larger starch molecules into a mix of fermentable and unfermentable sugars. This rest results in a fuller-bodied beer with more residual sweetness.

Benefits of Step Mashing:

- Improved Efficiency: Optimizes enzyme activity, leading to more complete starch conversion and higher extraction efficiency—the effectiveness of transferring desired compounds into the wort.
- Control Over Wort Fermentability: Adjusting the temperature and duration of each rest allows brewers to fine-tune the balance of fermentable and unfermentable sugars.
- Enhanced Flavor Development: Different enzymes contribute to the development of various flavor compounds, allowing for greater complexity in the finished beer.

Step Mashing (Kettle Only):

1. Preheat the Insulated Mash Tun: Preheat with 180°F (82°C) water for 20 minutes, then return the water to the kettle or hot liquor tank.
2. Fill Mash Tun with Strike Water: Fill with water at a strike temperature 9-10°F (5-6°C) higher than the desired mash temperature, e.g., 131°F (55°C).
3. Mash-In Grist: Add the grist, correcting the temperature to 122-131°F (50-55°C).
4. Correct Mash pH: Adjust pH to 5.2-5.6 using calcium carbonate, calcium chloride, or gypsum as needed.

5. Hold Temperature for Protein Rest: Hold for 30 minutes.
6. Ramp to Beta-Amylase Rest: Increase temperature to 140°F (60°C) at 1-2°F per minute. Hold for 15 minutes.
7. Ramp to Alpha-Amylase Rest: Increase to 150-155°F (66-68°C). Hold for 1-1½ hours.
8. Mash-Out: Boost temperature to 165-168°F (74-76°C). Hold for 5 minutes.

Double Mashing

Double mashing involves using two separate mashes: one with adjuncts and the other with the main malt. This technique is often used when brewing with adjuncts like rice or corn, which require gelatinization—the process of heating starches to make them soluble and accessible to enzymatic breakdown during mashing.

Double Mashing (Kettle Only):

1. Prepare Adjunct Mash:
    - Preheat Mash Tun: Preheat with 180°F (82°C) water for 20 minutes.
    - Mash-In Adjuncts: Fill with 4 gallons of water for every 1-1½ pounds of adjunct grist (rice or corn) with 10-15% crushed pale malt, correcting temperature to 155-158°F (68-70°C).
    - Correct pH: Adjust to 5.2-5.6.
    - Hold Temperature: Hold for 20 minutes.
    - Boil until Gelatinized: Boil for about 20 minutes.
2. Prepare Main Mash:
    - Mash-In Main Malt: Mix remaining crushed malt with 1 gallon of water for every 4 pounds of grist to an end temperature of 113°F (45°C). Hold for an extended protein rest.

3. Combine Mashes:
    - Add Adjunct Mash to Main Mash: Resulting in a temperature of 155°F (68°C).
    - Hold Temperature: Until starch conversion is complete. Use the simple and quick iodine test to check for residual starch. If the solution turns blue or black, it indicates that starches are still present and have not been fully converted to sugars.
4. Mash-Out: Boost temperature to 165-168°F (74-76°C). Hold for 5 minutes.

Decoction Mashing

Decoction mashing is a traditional technique where a portion of the mash is removed, boiled, and then returned to the main mash. This method, common in traditional German lagers like Pilsner, Bock, and Märzen, enhances malt character and improves extraction. The boiling process creates Maillard reactions—chemical processes that occur between amino acids and reducing sugars, contributing to the development of complex flavors, aromas, and colors in malt and, consequently, in beer. These reactions are responsible for creating toasty, nutty, and caramel-like flavors.

Process Overview:

1. Dough-In: Mix the malt with water to create a thick mash.
2. Decoction Pull: Remove a portion of the thick mash (maximizing solids) and boil it for 20 minutes—this gelatinizes the starches and develops flavors through Maillard reactions.
3. Return and Mix: Add the boiled decoction back to the main mash, raising the overall temperature and activating different enzymes.
4. Repeat: Multiple decoction pulls can be performed to achieve the desired mash profile.

Classic Triple-Decoction:

1. Preheat Mash Tun: With 180°F (82°C) water for 20 minutes.
2. Dough-In Grist: Mix with a small amount of cold water.
3. Infuse Boiling Water: Until 95°F (35°C) for an acid rest.
4. Correct pH: Adjust to 5.2-5.6.
5. First Decoction: Remove about 1/3 of the mash, boil for 20 minutes, return to Mash Tun to raise temperature to 122°F (50°C) for protein rest.
6. Second Decoction: Remove portion, boil, return to raise temperature to 150°F (66°C). Hold for 15 minutes. Use the simple and quick iodine test to check for residual starch. If the solution turns blue or black, it indicates that starches are still present and have not been fully converted to sugars..
7. Mash-Out Decoction: Remove portion, boil, return to raise temperature to 168°F (76°C). Hold for 5 minutes.

Challenges in Mashing

Advanced mashing techniques present several challenges, including maintaining temperature accuracy, managing pH stability, and preventing stuck mashes.

Water-to-Grain Ratio and Mash Stiffness

The water-to-grain ratio affects the mash's consistency and efficiency. Mash stiffness refers to this ratio, affecting the mash's thickness or viscosity.

- A stiffer mash (around 1.33 quarts of water per pound of grain) favors initial starch and protein breakdown, improving wort clarity but may inhibit further enzyme activity due to high sugar concentration.
- A thinner mash (approximately 2.5 quarts per pound) allows more complete carbohydrate breakdown but takes longer and can cause faster enzyme deactivation.

Adjusting this ratio can influence the thickness of the mash and the effectiveness of starch conversion. Proper water-to-grain ratio ensures optimal enzyme activity and efficient extraction of fermentable sugars.

Temperature Control

Use precise thermometers and regularly calibrate equipment to ensure accurate temperature control. Temperature accuracy is crucial for activating the correct enzymes and achieving the desired mash profile. Use temperature-controlled Mash Tuns and precise heating methods to avoid fluctuations. This accuracy is essential for maintaining the quality and consistency of the beer.

Mash pH Management

Maintaining mash pH between 5.2 and 5.6 is crucial for optimal enzyme activity and flavor stability. Regularly check and adjust pH using food-grade acids or bases as needed. Proper pH management ensures efficient starch conversion and consistent beer quality, preventing off-flavors and enhancing the beer's overall stability.

Stirring and Uniformity

Regularly stir the mash to ensure even temperature distribution and prevent the formation of hot spots. Consistent stirring helps maintain uniform enzyme activity throughout the mash, leading to better starch conversion and a more consistent final product.

Preventing Stuck Mash

A stuck mash occurs when the mash becomes compacted, hindering the flow of wort during lautering and sparging. Sparging is the process of rinsing the grain bed with hot water during lautering to extract the remaining sugars from the mash.

- Proper Grain Crush: Ensure grains are crushed appropriately —not too fine—to prevent compaction.
- Grain Bed Depth: Avoid overly deep grain beds that can compact under their weight.

- Use of Rice Hulls: Adding rice hulls—an adjunct that increases porosity—can prevent stuck sparges by creating a more porous grain bed.

Conclusion

The mashing process is a critical and intricate step in brewing that transforms raw ingredients into a fermentable solution, setting the foundation for the beer's final character. By understanding the roles of enzymes, the importance of temperature and pH control, and the different mashing methods, brewers can craft a wide variety of beer styles with precision and creativity.

This chapter's exploration of mashing connects deeply with the themes of rationalism, empiricism, and the scientific method, showcasing how a blend of theoretical knowledge and empirical practice leads to brewing success. Mastery of mashing techniques allows brewers to manipulate the beer's body, flavor, and alcohol content, creating unique and high-quality brews.

> *To mash with care is to honor the unseen labor of nature—where precision becomes virtue, and even enzymes are owed our respect.*

**

Review Questions

True/False Questions

1. True or False: Mashing involves combining crushed malt with hot water to create a mash, where enzymes break down starches into fermentable sugars.

2. True or False: The typical pH for a mash should be between 4.2 and 4.6 to ensure optimal enzyme activity.

3. True or False: Alpha-amylase operates optimally at higher temperatures and breaks down complex starches into fermentable sugars and dextrins.

4. True or False: Step mashing involves maintaining a constant temperature throughout the mashing process.

5. True or False: Decoction mashing is a traditional technique where a portion of the mash is removed, boiled, and then returned to the main mash.

Multiple Choice Questions

6. Which enzyme is most effective at lower temperatures and produces maltose, a fermentable sugar?
A) Protease
B) Alpha-amylase
C) Beta-amylase
D) Glucanase

7. What is the purpose of a protein rest at 122°F during the mashing process?
A) To ferment the beer
B) To enhance hop bitterness
C) To break down proteins into amino acids and peptides
D) To boil the wort

Brewer Vignettes

8. Brewer Vignette 1: Imagine you are a brewer aiming to create a fuller-bodied beer. You hypothesize that using a higher mashing temperature will help achieve this. Describe the steps you would take to test this hypothesis and the expected results.
A) Use random mashing temperatures without monitoring.
B) Avoid measuring the temperature and focus on pH adjustments.
C) Maintain a higher mashing temperature (158-162°F), monitor enzyme activity, and measure the resulting sugar profile.
D) Use a constant low temperature for the entire mashing process.

9. Brewer Vignette 2: As a brewer, you want to enhance the malt character of your German-style lager using decoction mashing. Explain how you would implement this technique and what benefits you expect.
A) Use only cold water throughout the mashing process.
B) Avoid boiling any part of the mash and rely solely on temperature rests.
C) Ignore decoction and use single infusion mashing.
D) Perform multiple decoction pulls, boil the pulled mash, and return it to the main mash to develop complex malt flavors.

10. Brewer Vignette 3: You are trying to avoid stuck mashes in your brewing process. Describe the methods you would use to prevent this issue and ensure a smooth mashing process.
A) Regularly stir the mash, use rice hulls to improve porosity, and ensure proper grain crush and bed depth.
B) Add more adjuncts without monitoring the mash consistency.
C) Use a very fine grain crush to create a compact mash.
D) Avoid using rice hulls and let the mash compact naturally.

**

Correct Answers
True/False Questions
1. True
2. False (The typical pH for a mash should be between 5.2 and 5.6.)
3. True
4. False (Step mashing involves gradually increasing the temperature to activate different enzymes at their optimal temperatures.)
5. True

Multiple Choice Questions
6. C) Beta-amylase
7. C) To break down proteins into amino acids and peptides

Brewer Vignettes
8. C) Maintain a higher mashing temperature (158-162°F), monitor enzyme activity, and measure the resulting sugar profile.
9. D) Perform multiple decoction pulls, boil the pulled mash, and return it to the main mash to develop complex malt flavors.
10. A) Regularly stir the mash, use rice hulls to improve porosity, and ensure proper grain crush and bed depth.

**

Beyond The Chapter

- Bamforth, C. W. (2003). Beer: Tap into the Art and Science of Brewing. Oxford University Press.
    - *Explains the science behind mashing and its role in the brewing process.*
- Palmer, J. (2006). How to Brew: Everything You Need To Know To Brew Beer Right The First Time. Brewers Publications.
    - *Covers the practical aspects of mashing and optimizing malt extraction.*
- Noonan, G. (1996). Brewing Lager Beer: The Most Comprehensive Book for Home - And Microbrewers. Brewers Publications.

    - *Provides detailed information on mashing techniques for brewing lagers.*
- Kunze, W. (2014). Technology Brewing & Malting. VLB Berlin.
    - *An in-depth technical resource on mashing and malt processing.*
- Daniels, R. (1996). Designing Great Beers: The Ultimate Guide to Brewing Classic Beer Styles. Brewers Publications.
    - *Discusses mashing protocols for creating various classic beer styles.*

Weblinks

- The Mashing Process - Brew Your Own: https://byo.com/?s=Mashing&post_type=article&orderby=relevance
- Understanding Mashing - Craft Beer & Brewing Magazine: https://beerandbrewing.com/search/?search=Mashing
- The Science of Mashing - American Homebrewers Association: https://www.homebrewersassociation.org/?s=Mashing
- Decoction Mashing - Brew Cabin: https://www.brewcabin.com/decoction-mash/
- Temperature and pH in Mashing - Homebrew Academy: https://homebrewacademy.com/education/beer/

These references provide deeper insights into the mashing process, offering valuable resources for both homebrewers and professionals.

***

Chapter 5: The Art of Lautering—Filtering Clarity and Quality from the Mash

Lautering is the art of precision, where clarity and quality are filtered from the depths of the mash.

Imagine the moment after mashing, where the sweet, fragrant wort needs to be carefully separated from the grain husks. This is lautering, a critical step in brewing that ensures the transfer of fermentable sugars while leaving behind the spent grains. The efficiency and effectiveness of this process can significantly impact the beer's clarity, flavor, and overall quality.

In brewing, lautering involves the separation of the liquid wort from the solid grain particles after mashing. Key concepts such as the separation of wort and grain, sparging methods, and the importance of preventing stuck sparges are central to understanding lautering. Mastery of this process is essential for both novice and expert brewers aiming to produce high-quality beer.

Ancient Origins

The origins of lautering can be traced back to ancient brewing practices. Early brewers discovered through empirical observation that separating the liquid from the grains resulted in a clearer, more consistent brew. In ancient Mesopotamia and Egypt, rudimentary lautering techniques involved simple filtration methods using woven baskets or reed mats. These early methods were primarily based on trial and error, with brewers learning from each batch and refining their processes over time.

These ancient brewers realized that leaving grains in the fermenting liquid led to undesirable flavors and textures. By removing the grains, they could produce a more palatable beverage. The leftover grain, known as spent grain, was often repurposed. Spent grain—the leftover grain from the brewing process—could be composted or used as animal feed, demonstrating an early understanding of sustainability and resourcefulness.

The basic principles of lautering were laid out in these early civilizations, forming the foundation for more advanced techniques.

Medieval Europe

During the medieval period, brewing techniques became more sophisticated. Monastic breweries in Europe played a crucial role in advancing lautering methods. Monks, with their meticulous record-keeping, documented every step in the brewing process to help track and analyze variables, identify patterns, and improve techniques. This systematic approach led to significant advancements in brewing.

The introduction of the lauter tun, a vessel specifically designed for separating wort from grain, marked a significant milestone. The lauter tun typically featured a false bottom—a perforated plate or screen at the bottom that allows wort to pass through while retaining the grain husks. This design facilitated effective separation and improved wort clarity.

The formation of the grain bed during lautering acted as a natural filter, further enhancing the clarity of the wort. The grain bed—the layer of crushed malt and adjuncts in the lauter tun during lautering—traps unwanted particles, ensuring a cleaner wort for fermentation.

These innovations improved the consistency and quality of the beer produced during this time. Monastic brewers' dedication to refining their techniques laid the groundwork for modern brewing practices.

The Industrial Revolution

The Industrial Revolution brought about significant technological advancements in brewing. The development of precision instruments like thermometers and hydrometers enabled brewers to monitor and control the lautering process with greater accuracy. The introduction of mechanical stirring devices helped maintain even temperature distribution and prevented the grain bed from compacting during lautering.

Advancements in equipment reduced the risk of a stuck sparge—a condition where the grain bed compacts, preventing wort from flowing through, often caused by improper grain crush or bed depth. By preventing compaction, mechanical stirrers ensured a consistent flow of wort, improving efficiency and quality.

These innovations were driven by both rationalism, through the application of scientific principles, and empiricism, through continual experimentation and refinement. This period marked the beginning of modern brewing techniques, combining scientific understanding with practical application.

Rationalism—Deductive Reasoning

Rationalism played a vital role in understanding the physical principles underlying lautering. Theoretical frameworks in fluid dynamics and filtration guided the development of efficient lautering techniques. Brewers applied top-down deductive reasoning to optimize the flow of wort through the grain bed and prevent issues such as channeling and stuck sparges.

Theoretical Frameworks

The application of Darcy's Law, which describes the flow of fluid through a porous medium, is an example of rationalism in lautering. Darcy's Law states that the flow rate through a porous medium is proportional to the pressure difference and inversely proportional to the medium's resistance. Understanding this principle allowed brewers to design lauter tuns and grain beds that facilitate efficient wort flow.

By adjusting factors such as grain crush size and bed depth, brewers optimized the lautering process. The development of sparging techniques, such as fly sparging and batch sparging, was also guided by theoretical knowledge of fluid mechanics and the behavior of liquids under different conditions.

- Fly Sparging: A continuous sparging method where sparge water is added to the mash tun while wort is simultaneously drained from the bottom, maximizing sugar extraction. This method requires careful monitoring to avoid channeling, where water flows through paths of least resistance in the grain bed, leaving pockets of unextracted sugars and reducing lautering efficiency.
- Batch Sparging: A sparging method where the wort is completely drained from the mash tun before sparge water is added. After the initial runoff, sparge water is added in a single batch, the grain bed is stirred to redistribute sugars, and then the wort is drained again. This method simplifies the process and reduces the risk of channeling since the entire grain bed is resaturated and mixed. Batch sparging is popular among homebrewers due to its simplicity and ease of use, even if it may be slightly less efficient in sugar extraction compared to fly sparging.

These theoretical frameworks provided a scientific basis for optimizing the lautering process, ensuring consistency and quality in the final product.

Empiricism—Inductive Reasoning

Refining Lautering Practices

Empiricism has been crucial in refining lautering practices. Brewers have relied on observations and experiments to enhance their understanding and optimize the process. Through inductive reasoning, they have developed practical solutions to common challenges encountered during lautering.

Observational Learning

Historical records show that brewers continually adjusted their lautering techniques based on observations. For example, they noticed that certain grain crush sizes or bed depths affected the efficiency of wort separation. If the grain was crushed too finely, it led to a compacted grain bed and a stuck sparge. Conversely, a coarser crush improved flow but could reduce sugar extraction efficiency.

They also observed that maintaining an even grain bed depth was essential for proper filtration without causing blockages. These observations led to empirical adjustments that improved the lautering process. By carefully observing and documenting their findings, brewers were able to refine their methods over time.

Experimental Results

Empirical data from experiments conducted by brewers have driven many advancements in lautering. For instance, the realization that maintaining a consistent sparge water temperature improves sugar extraction without extracting undesirable tannins was a significant empirical finding.

- Tannins are naturally occurring polyphenolic compounds found in barley husks, hops, and certain types of wood. They can contribute to the astringency and bitterness of beer. During the brewing process, excessive extraction of tannins, often due to high temperatures or improper pH levels during mashing and sparging, can result in unwanted harsh flavors and haze formation.

By controlling sparge water temperature and pH, brewers minimized tannin extraction, leading to a smoother, more palatable beer. These experimental practices have shaped modern lautering techniques, demonstrating the importance of inductive reasoning in brewing.

The Scientific Method—Advancing Lautering Techniques

Application of the Scientific Method

The scientific method has been instrumental in advancing lautering techniques. This approach involves systematic observation, hypothesis formation, experimentation for verifying or falsifying the hypothesis, data analysis with assessment, replicability, and peer review. By applying the scientific method, brewers have developed and refined their lautering techniques based on solid evidence and rigorous testing.

Case Study: Development of Fly Sparging

One notable example is the development of fly sparging. Brewers hypothesized that continuously adding sparge water while simultaneously draining wort would maximize sugar extraction. They designed experiments to test variables such as flow rate, water temperature, and pH.

- Experimentation: Brewers conducted trials comparing fly sparging with other methods, measuring sugar extraction efficiency and wort clarity.
- Data Analysis: Results showed that fly sparging increased extraction efficiency but required careful control to prevent channeling.
- Replicability and Peer Review: Findings were shared within the brewing community, leading to widespread adoption and further refinement of the technique.

This collaborative approach exemplifies the power of the scientific method in solving practical brewing challenges.

Separation of Wort and Grain

The Lautering Process

Lautering begins in the lauter tun, where the liquid wort is separated from the solid grain husks. The lauter tun features a false bottom or slotted plate that allows the liquid wort to pass through while retaining the solid grain husks.

The grain bed acts as a natural filter, trapping unwanted particles and clarifying the wort. Proper formation and maintenance of the grain bed are essential for effective filtration.

Importance of Separation

The separation process is crucial for ensuring all fermentable sugars are extracted from the grains. A well-executed lautering process contributes to the clarity and quality of the final beer by minimizing the presence of unwanted grain particles and ensuring that the wort is free from haze-causing compounds.

Effective separation also prevents the extraction of excessive proteins and lipids that can cause fermentation disorders, poor head retention, and off-flavors. By optimizing the separation of wort and grain, brewers lay the foundation for producing high-quality beer.

Sparging

Once starch conversion is complete, the draff—the solid part of the mash consisting of husks, small particles, and flour-like fines—must be separated from the liquid part of the mash—the sweet wort—before boiling. Sparging is the technique used to rinse the grains with hot water, extracting the remaining sugars from the mash.

There are two main sparging methods:

1. Fly Sparging: As previously described, this method involves the continuous addition of sparge water while simultaneously draining the wort. It maximizes sugar extraction but requires careful monitoring to avoid channeling.
2. Batch Sparging: As previously described, is a lautering method where wort is completely drained from the mash tun, sparge water is added in a single batch, stirred, and then drained again. It is popular among homebrewers for its simplicity and ease of use.

Fly Sparging Procedure

- Sparge Water Preparation:
  - Heat sparge water to 163°F (73°C) if mash-out is performed, or to 168°F (75°C) if no mash-out is performed.
  - Adjust mash pH to 5.7 or below. Acidify sparge water with lactic or phosphoric acid if necessary.
- Transfer Mash:
  - Transfer mash to the lauter tun and let it stand for 15 minutes to settle.
- Vorlauf (Recirculation):
  - Recirculate wort at a rate of ½ quart per minute for 15 minutes to clarify. If the wort clears too quickly, draining is too fast; if it clears too slowly, draining is too slow.
  - Ensure the pH of runoff is at or below 5.7, adjusting if necessary with food-grade phosphoric acid.
- Runoff and Sparging:
  - Begin runoff, keeping the wort level 1-2 inches above the grain bed to prevent compaction.
  - Add sparge water at the same rate as wort outflow, maintaining the water level above the grain bed.
  - Sparge for approximately 45 minutes, with the rate increasing over time.
- Completion:
  - Recirculation, runoff, and sparging take about 1½ to 2 hours to complete.

Batch Sparging Procedure

- Stir and Rest:
  - After mashing, stir the mash thoroughly and let it rest for 10-15 minutes.

- First Runoff:
  - Vorlauf: Recirculate until the wort runs clear, then drain completely into the boil kettle.
- Add Sparge Water:
  - Add sparge water heated to 168°F (75°C) into the mash tun.
- Stir and Rest:
  - Stir the mash thoroughly and let it rest for 10-15 minutes.
- Second Runoff:
  - Vorlauf: Recirculate until the wort runs clear, then drain completely into the boil kettle.

Batch sparging is less complex and requires less equipment than fly sparging, making it accessible for homebrewers. While it may be slightly less efficient, it offers greater simplicity and ease of use.

Challenges

Stuck Sparge

A stuck sparge occurs when the grain bed compacts, preventing the wort from flowing through. This issue can significantly disrupt the brewing process, leading to inefficient extraction and potential quality issues in the final beer.

Prevention Strategies:

- Proper Grain Crush: Ensure grains are crushed correctly, breaking them but leaving husks mostly intact to create a porous grain bed.
- Grain Bed Depth: Maintain an even grain bed depth to facilitate good wort flow without causing blockages.
- Use of Rice Hulls: Adding rice hulls can increase porosity and prevent stuck sparges by creating a more porous grain bed.

Sparge Water Temperature

Keep sparge water at around 168°F (75°C) to improve sugar extraction without extracting tannins. Maintaining the correct temperature helps maximize sugar extraction while minimizing the risk of undesirable compounds entering the wort.

pH Control

Maintain sparge water pH below 6 to prevent tannin extraction. Acidify sparge water with lactic or phosphoric acid if necessary. Proper pH control ensures that the wort remains balanced and free from harsh, astringent notes.

Evidence-Based Solutions

Brewers often encounter challenges during the lautering process, such as maintaining the correct flow rate, preventing stuck sparges, and achieving the desired wort clarity—the absence of grain particles and other solids in the wort.

Solutions:

- Flow Control: Use flow meters or manual adjustments to control the rate of wort runoff and sparge water addition.
- Monitoring: Regularly monitor the flow and be prepared to gently stir the grain bed if signs of compaction occur.
- Vorlauf (Recirculation): Perform a thorough vorlauf to clarify the wort before collecting it in the boil kettle.

Implementing these solutions helps brewers overcome common challenges and improve their lautering process, leading to higher-quality beer.

Conclusion

Lautering, with its meticulous separation of wort and grain, embodies the themes of rationalism, empiricism, and the scientific method. Rationalism is evident in understanding the physical principles governing fluid dynamics and filtration. Empiricism is reflected in the trial and error, observations, and adjustments made by brewers over centuries to perfect their techniques.

The scientific method, through observation, hypothesis formation, experimentation, data analysis, replicability, and peer review, continues to drive innovations in lautering, ensuring that brewers can consistently produce high-quality beer. By integrating these approaches, brewers can optimize their lautering process and produce exceptional beers.

> *To lauter is to practice clarity as a virtue—discerning what to keep, what to let go, and doing so with patience, purpose, and care. It is the brewer's quiet lesson in letting the good pass through.*

**

REVIEW QUESTIONS

True/False Questions

1. True or False: Lautering is the process of separating the liquid wort from the solid grain husks after mashing.

2. True or False: Fly sparging involves draining the wort completely from the mash tun, then adding a batch of sparge water, stirring, and draining again.

3. True or False: The pH of sparge water should be maintained below 6 to prevent tannin extraction.

4. True or False: The use of false bottoms or slotted plates in lauter tuns was a significant innovation during the medieval period.

5. True or False: Empiricism has been crucial in refining lautering practices through observations and experiments.

Multiple Choice Questions

6. Which sparging method involves the continuous addition of sparge water to the mash tun while wort is simultaneously drained from the bottom?
A) Batch sparging
B) Fly sparging
C) Single infusion sparging
D) Double decoction sparging

7. What is the recommended temperature for sparge water to improve sugar extraction without extracting tannins?
A) 148°F (64°C)
B) 168°F (75°C)
C) 158°F (70°C)
D) 178°F (81°C)

Brewer Vignettes

8. Brewer Vignette 1: Imagine you are a brewer aiming to prevent stuck sparges during lautering. Describe the steps you would take to ensure a smooth lautering process.
A) Use a very fine grain crush to create a compact mash.
B) Regularly stir the mash, use rice hulls to improve porosity, and ensure proper grain crush and bed depth.
C) Add more adjuncts without monitoring the mash consistency.
D) Avoid using rice hulls and let the mash compact naturally.

9. Brewer Vignette 2: As a brewer, you want to maximize sugar extraction using fly sparging. Explain how you would implement this technique and what benefits you expect.
A) Drain the wort completely before adding sparge water.
B) Continuously add sparge water while draining wort, monitoring to avoid channeling, for efficient sugar extraction.
C) Use cold water for sparging.
D) Avoid sparging and rely solely on the initial mash.

10. Brewer Vignette 3: You are noticing that your beer has astringent flavors, possibly due to tannin extraction during sparging. Describe how you would adjust your process to avoid this issue.
A) Use hotter sparge water.
B) Maintain sparge water at 168°F and adjust pH to below 6 using lactic or phosphoric acid.
C) Increase the grain crush size.
D) Sparge with cold water only.

**

Correct Answers
True/False Questions
1. True
2. False (Fly sparging involves the continuous addition of sparge water to the mash tun while wort is simultaneously drained from the bottom.)
3. True
4. True
5. True

Multiple Choice Questions
6. B) Fly sparging
7. B) 168°F (75°C)

Brewer Vignettes
8. B) Regularly stir the mash, use rice hulls to improve porosity, and ensure proper grain crush and bed depth.
9. B) Continuously add sparge water while draining wort, monitoring to avoid channeling, for efficient sugar extraction.
10. B) Maintain sparge water at 168°F and adjust pH to below 6 using lactic or phosphoric acid.

**

BEYOND THE CHAPTER

- Bamforth, C. W. (2003). Beer: Tap into the Art and Science of Brewing. Oxford University Press.
  - *Examines the lautering process and its significance in brewing.*
- Palmer, J. (2006). How to Brew: Everything You Need To Know To Brew Beer Right The First Time. Brewers Publications.
  - *Provides practical tips for efficient lautering and maximizing wort clarity.*
- Kunze, W. (2014). Technology Brewing & Malting. VLB Berlin.

- *Details the technical aspects of lautering and modern equipment used.*
- Noonan, G. (1996). Brewing Lager Beer: The Most Comprehensive Book for Home - And Microbrewers. Brewers Publications.
  - *Discusses traditional and modern lautering techniques for lager production.*
- Daniels, R. (1996). Designing Great Beers: The Ultimate Guide to Brewing Classic Beer Styles. Brewers Publications.
  - *Explores lautering methods used for different beer styles and their effects on wort quality.*

Weblinks

- The Lautering Process - Brew Your Own: https://byo.com/?s=Lautering&post_type=article&orderby=relevance
- Lautering in Homebrewing - Craft Beer & Brewing Magazine: https://beerandbrewing.com/search/?search=Lautering
- The Science of Lautering - American Homebrewers Association: https://www.homebrewersassociation.org/?s=Lauterings
- Efficient Lautering Practices - Brew Cabin: https://www.brewcabin.com/homebrewing/

These references provide deeper insights into lautering techniques, offering valuable resources for both homebrewers and professionals.

***

Chapter 6: The Transformative Power of Boiling—Forging the Heart and Soul of Every Great Brew

Boiling transforms wort with precision and care, forging the heart and soul of every great brew.

Imagine a bustling brewery, the air thick with the aromatic steam rising from boiling kettles. This stage—the boiling of the wort—is where many of the most critical transformations in brewing occur. It is a moment where science and craft converge, ensuring the beer's quality, flavor, and stability. Boiling serves multiple essential functions: it sterilizes the wort, extracts bitterness and aroma from hops, facilitates protein coagulation, and concentrates the wort. These processes are pivotal in defining the beer's final character. The boiling stage is not just a routine procedure but a meticulously controlled process rooted in rationalism, empiricism, and the scientific method.

Ancient Origins

The practice of boiling wort dates back to ancient brewing traditions. Early brewers in Mesopotamia and Egypt discovered that boiling wort helped preserve the quality and safety of beer. These early practices were based on empirical observations—boiled wort resulted in better-tasting and more stable beer, likely due to the elimination of harmful microorganisms. While they may not have understood the concept of sterilization, they recognized that heat treatment prevented spoilage and extended the beer's shelf life. This empirical knowledge, though not fully understood in scientific terms, laid the groundwork for future brewing techniques.

Medieval Europe

During the medieval period, brewing techniques evolved significantly in European monasteries. Monks documented their brewing methods meticulously, contributing to the refinement of boiling practices. The introduction of copper kettles improved heat transfer and allowed for more consistent boiling temperatures. Copper is an excellent conductor of heat, facilitating efficient heat transfer—the process of transferring heat to the wort during boiling, affecting evaporation, isomerization of alpha acids, and protein coagulation.

The use of hops marked a significant milestone. Monks began adding hops to the boiling wort, discovering that hops not only added bitterness but also acted as a preservative, enhancing the beer's stability. The practice of hop additions—adding hops to the boiling wort at different times to achieve desired levels of bitterness, flavor, and aroma—became an integral part of brewing.

This period also saw the application of rationalism, as brewers began to understand and control the boiling process more scientifically. The combination of empirical observation and rational analysis led to more consistent and higher-quality beer.

The Industrial Revolution

The Industrial Revolution brought technological innovations that transformed boiling practices. Precision instruments like thermometers and hydrometers enabled brewers to monitor and control the boiling process with greater accuracy. The development of steam-powered brewing systems further improved efficiency and consistency.

Brewers began to understand the importance of boil duration—the length of time the wort is boiled, typically ranging from 60 to 90 minutes—which affects wort concentration, hop utilization, and flavor development. Longer boils could enhance wort concentration and hop utilization but also risked caramelization and unwanted flavors.

Modern Era

Modern brewing continues to advance, driven by both rationalism and empiricism, with brewers applying scientific principles and refining techniques through experimentation. Today's brewers have a deep understanding of the chemical and physical changes that occur during boiling, allowing them to manipulate variables to achieve desired outcomes.

For example, precise control over evaporation rate—the rate at which water evaporates from the wort during boiling—affects wort concentration and final gravity. Consistent evaporation is essential for achieving desired beer characteristics.

Rationalism—Deductive Reasoning

Heat Transfer and Chemical Reactions

Rationalism has been crucial in understanding the physical and chemical principles underlying the boiling process. By applying top-down deductive reasoning, brewers have studied heat transfer, evaporation, and chemical reactions to develop efficient boiling techniques. This understanding allows for precise control over the brewing process, ensuring consistency and quality.

Theoretical Frameworks

The application of principles such as thermodynamics and chemical kinetics has guided the development of boiling practices. Understanding how heat causes the evaporation of water and the isomerization of alpha acids in hops has allowed brewers to optimize the boiling process.

- Isomerization is the chemical process by which alpha acids in hops are converted to iso-alpha acids during boiling, imparting bitterness to the beer.

Brewers recognize that the rate of isomerization depends on factors such as temperature, pH, and boil duration. By manipulating these variables, they can control the level of bitterness in the final product.

Theoretical frameworks have informed decisions about boil duration, temperature control, and hop addition timing, ensuring that each batch of beer meets the desired specifications. These frameworks provide a scientific basis for refining boiling techniques and achieving the desired outcomes.

Empiricism—Inductive Reasoning

Refining Boiling Practices

Empiricism, through observations and experiments, has played an equally important role in refining boiling practices. Brewers have used bottom-up inductive reasoning to gather empirical data and shape their understanding of the boiling process. This approach allows for continuous improvement based on real-world results.

Observational Learning

Historical records show that brewers continually adjusted their boiling techniques based on observations. They noted that certain boil times or hop addition schedules affected the beer's flavor and stability.

For example, experiments revealed that adding hops at different times during the boil impacted the beer's bitterness, flavor, and aroma. These observations led to the categorization of hops into bittering hops, flavor hops, and aroma hops, each added at specific times to achieve desired effects.

Experimental Results

Empirical data from experiments conducted by brewers have driven many advancements. For instance, experiments on hop utilization—the efficiency with which alpha acids in hops are isomerized during boiling—revealed how different hop varieties and addition times influence bitterness and aroma.

- Hop Utilization is influenced by boil duration, temperature, and wort composition. Longer boil times generally increase hop utilization, leading to higher bitterness.

By experimenting with different variables, brewers have been able to refine their techniques and enhance the quality of their beer.

The Scientific Method—Advancing Boiling Techniques

Application of the Scientific Method

The scientific method has been instrumental in advancing boiling techniques. This approach involves systematic observation, hypothesis formation, experimentation to verify or falsify the hypothesis, data analysis with assessment, replicability, and peer review. Applying the scientific method ensures that brewing practices are continually improved based on rigorous testing and analysis.

Case Study: Protein Coagulation (Hot Break)

One notable example is the study of protein coagulation, also known as the hot break. Brewers hypothesized that a vigorous boil would cause proteins and tannins to coagulate and precipitate out of the wort.

- Protein Coagulation (Hot Break): During the boiling phase, heat causes water molecules to detach from larger proteins in the wort, a process known as denaturing. This makes the large protein molecules insoluble, leading them to flocculate (clump together). Vigorous boiling aids in the flocculation of proteins, tannins (polyphenols), and lipids, resulting in the "hot break." This separation is crucial for the clarity and stability of the final beer. However, achieving a hot break is challenging if the wort's pH is lower than 5.0.

Experiments confirmed this hypothesis, showing that effective protein coagulation improves beer clarity and stability. Data analysis and peer review within the brewing community led to widespread adoption and further refinement of this technique.

Four Purposes of Boiling

1. Sterilization

The primary purpose of boiling is to sterilize the wort by killing unwanted microorganisms. The high temperatures achieved during boiling ensure that harmful bacteria and wild yeast are eliminated, preventing contamination and ensuring a controlled and predictable fermentation process.

- Sterilization refers to the process of eliminating all microorganisms from equipment, containers, and surfaces to prevent contamination. Proper sterilization ensures that only the desired yeast strains ferment the wort, resulting in a clean and uncontaminated beer.
- Wild Yeast are naturally occurring yeast strains that can contribute unique flavors and characteristics to beer. However, in most brewing processes, wild yeast are considered contaminants that can cause off-flavors or unpredictable fermentation results.

Effective sterilization is essential for producing consistent, high-quality beer.

2. Hop Utilization

Boiling is crucial for hop utilization, extracting bitterness, flavor, and aroma compounds from hops. The chemical process of isomerization of alpha acids during boiling imparts the desired bitterness to the beer.

- Alpha Acids are compounds in hops that contribute to bitterness in beer through isomerization during boiling. They are less affected by dry hopping, which focuses on aroma and flavor.

The timing of hop additions during the boil significantly influences the beer's final characteristics, allowing brewers to create a balanced and complex flavor profile. Proper hop utilization is key to achieving the desired bitterness and aroma in the final product.

3. Protein Coagulation (Hot Break)

Protein coagulation, or the hot break, occurs during boiling and is essential for improving beer clarity and stability. The heat causes proteins and tannins to coagulate and form clumps, which are then removed from the wort.

This process helps reduce haze and enhances the overall appearance and shelf stability of the beer. Effective protein coagulation is crucial for producing clear and stable beer.

4. Wort Concentration

Boiling also concentrates the wort by evaporating excess water, which increases the wort's gravity. This concentration step is crucial for controlling the beer's final alcohol content and body.

- Wort Concentration: The process of evaporating excess water during boiling to increase the wort's gravity, crucial for controlling the beer's final alcohol content and body. The wort pH should be between 5.0 and 5.4 to avoid harsh bitterness and achieve the best flavor from the hops during the boil.

By carefully monitoring the evaporation rate, brewers can achieve the desired wort concentration, ensuring consistency in the final product.

Three Types of Hop Additions

1. Bittering Hops

Bittering hops are added at the beginning of the boil and contribute primarily to the beer's bitterness. The long boil time, typically 60 to 90 minutes, ensures maximum isomerization of alpha acids, which impart the desired bitterness to the beer.

Practical Applications:

- Select hops with high alpha acid content for efficient bitterness.
- Add bittering hops at the start of the boil to maximize isomerization.

- Monitor boil duration to control the level of bitterness.

Understanding the role of bittering hops is essential for creating a well-balanced beer.

2. Flavor Hops

Flavor hops are added midway through the boil, usually between 15 and 30 minutes before the end, and contribute both bitterness and flavor. The shorter boil time preserves more of the hop's essential oils, adding complex flavors to the beer.

- Essential Oils: Volatile compounds in hops, such as myrcene, humulene, and caryophyllene, that contribute to aroma and flavor. They are best preserved through shorter boil times and dry hopping.

Practical Applications:

- Choose hops known for their desirable flavor profiles.
- Add flavor hops at the appropriate time to balance bitterness and flavor.
- Adjust addition timing to emphasize specific flavor characteristics.

The addition of flavor hops allows brewers to create a nuanced and complex taste.

3. Aroma Hops

Aroma hops are added at the end of the boil or during whirlpooling and impart intense hop aromas. The minimal exposure to heat preserves volatile aromatic compounds, enhancing the beer's nose.

- Whirlpooling: A technique used at the end of the boil where the wort is stirred to create a vortex, helping to separate and remove trub (solid particles) from the wort before fermentation. It also allows for hop additions that maximize aroma without increasing bitterness.

Practical Applications:

- Select hops with strong aromatic qualities.

- Add aroma hops during whirlpooling or as late additions to preserve essential oils.
- Consider dry hopping (adding hops during fermentation) for additional aroma enhancement.

Aroma hops are key to creating an appealing and aromatic beer.

Challenges

Boil Duration

The typical boil duration ranges from 60 to 90 minutes, depending on the desired beer characteristics. Longer boils can enhance wort concentration and hop utilization but may also increase the risk of caramelization and unwanted flavors.

Considerations:

- Monitor boil intensity to prevent excessive evaporation or scorching.
- Adjust boil time based on the beer style and desired characteristics.
- Be aware that longer boils may darken the wort and alter flavor profiles.

Evaporation Rate

Maintaining a consistent evaporation rate is crucial for controlling wort concentration and final gravity. Using precise equipment and monitoring tools helps brewers achieve the desired evaporation rate.

Strategies:

- Use calibrated equipment to measure volume before and after boiling.
- Control boil vigor to maintain consistent evaporation.
- Adjust heating systems as needed based on environmental conditions.

Protein Coagulation and pH Control

Achieving effective protein coagulation requires careful control of wort pH and boiling conditions.

- Maintain wort pH between 5.0 and 5.4 to avoid harsh bitterness and achieve the best flavor from the hops.
- Monitor and adjust pH using food-grade acids or bases as necessary.

Hop Utilization Consistency

Brewers often encounter challenges in achieving consistent hop utilization due to variations in hop alpha acid content and brewing conditions.

Solutions:

- Use standardized hop products with known alpha acid percentages.
- Adjust hop quantities based on alpha acid content and desired bitterness.
- Keep detailed records of hop additions and results for future reference.

Evidence-Based Solutions

Implementing evidence-based solutions helps brewers overcome common challenges and improve their boiling process.

- Use Calibrated Instruments: Employ accurate thermometers, hydrometers, and flow meters to monitor brewing parameters.
- Regular Testing: Analyze wort samples for gravity, pH, and other key indicators.
- Continuous Research: Stay informed about brewing science advancements and incorporate new techniques as appropriate.
- Peer Collaboration: Engage with the brewing community to share experiences and learn from others.

Conclusion

Boiling the wort is a critical stage in brewing that embodies the themes of rationalism, empiricism, and the scientific method. Rationalism is evident in understanding the physical and chemical principles governing boiling. Empiricism is reflected in the trial and error, observations, and adjustments made by brewers over centuries to perfect their techniques.

The scientific method, through observation, hypothesis formation, experimentation, data analysis, replicability, and peer review, continues to drive innovations in boiling, ensuring that brewers can consistently produce high-quality beer. By integrating these approaches, brewers can optimize their boiling process and produce exceptional beers.

> *To boil is to bear witness to transformation under fire—where clarity, strength, and character emerge not despite the heat, but because of it. In brewing, as in life, what we become depends on how we endure the boil.*

**

REVIEW QUESTIONS

True/False Questions

1. True or False: Boiling the wort helps sterilize it by killing unwanted microorganisms.

2. True or False: Empirical observations during the medieval period led to the introduction of hops in boiling wort.

3. True or False: The primary function of bittering hops is to contribute to the beer's aroma.

4. True or False: Protein coagulation during boiling is also known as the "*hot break*."

5. True or False: Maintaining a consistent evaporation rate during boiling is not important for wort concentration.

Multiple Choice Questions

6. Which stage of boiling wort primarily extracts bitterness from hops?
A) Aroma hop addition
B) Bittering hop addition
C) Flavor hop addition
D) Whirlpool hop addition

7. What is the typical duration for boiling wort in brewing?
A) 30 to 45 minutes
B) 60 to 90 minutes
C) 90 to 120 minutes
D) 120 to 150 minutes

Brewer Vignettes

8. Brewer Vignette 1: Imagine you are a brewer aiming to maximize hop aroma in your beer. Describe how you would adjust your boiling process to achieve this.
A) Add hops at the beginning of the boil.
B) Add hops midway through the boil.
C) Add hops at the end of the boil or during whirlpooling.
D) Avoid adding hops during the boil.

9. Brewer Vignette 2: As a brewer, you want to improve the clarity and stability of your beer. Explain how you would optimize protein coagulation during boiling.
A) Avoid boiling the wort vigorously.
B) Ensure a vigorous boil to facilitate protein coagulation, known as the hot break, and remove coagulated proteins.
C) Boil the wort at a low temperature.
D) Skip the boiling stage entirely.

10. Brewer Vignette 3: You are noticing inconsistent hop bitterness in your beer. Describe the methods you would use to control hop utilization during the boil.
A) Use random hop addition times without monitoring.
B) Calibrate thermometers and flow meters to monitor boil temperature and hop addition times precisely.
C) Add all hops at the end of the boil.
D) Avoid using any hops during the boil.

**

Correct Answers
True/False Questions
1. True
2. True
3. False (The primary function of bittering hops is to contribute to the beer's bitterness.)
4. True
5. False (Maintaining a consistent evaporation rate is crucial for wort concentration.)

Multiple Choice Questions
6. B) Bittering hop addition
7. B) 60 to 90 minutes

Brewer Vignettes
8. C) Add hops at the end of the boil or during whirlpooling.
9. B) Ensure a vigorous boil to facilitate protein coagulation, known as the hot break, and remove coagulated proteins.
10. B) Calibrate thermometers and flow meters to monitor boil temperature and hop addition times precisely.

**

Beyond The Chapter

- Bamforth, C. W. (2003). Beer: Tap into the Art and Science of Brewing. Oxford University Press.

  - *Explores the role of the boiling process in brewing and its impact on beer stability and flavor.*
- Palmer, J. (2006). How to Brew: Everything You Need To Know To Brew Beer Right The First Time. Brewers Publications.
  - *Provides practical guidance on the boiling process and its importance in brewing.*
- Kunze, W. (2014). Technology Brewing & Malting. VLB Berlin.
  - *Offers detailed information on the boiling process and modern equipment used.*
- Noonan, G. (1996). Brewing Lager Beer: The Most Comprehensive Book for Home - And Microbrewers. Brewers Publications.
  - *Discusses boiling techniques specific to lager production.*
- Daniels, R. (1996). Designing Great Beers: The Ultimate Guide to Brewing Classic Beer Styles. Brewers Publications.
  - *Examines boiling methods used for different beer styles and their effects on hop utilization.*

Weblinks

- Understanding Boiling - Brew Your Own: https://byo.com/article/boiling/
- The Science of Boiling - Craft Beer & Brewing Magazine: https://beerandbrewing.com/search/?search=boiling
- Advanced Boiling Methods - American Homebrewers Association: https://www.homebrewersassociation.org/?s=boiling

These references provide deeper insights into boiling techniques, offering valuable resources for both homebrewers and professionals.

***

Chapter 7: The Alchemy of Fermentation —Where Yeast Breathes Life into Wort

> *Fermentation is where yeast breathes life into wort, blending science and artistry into every brew.*

Imagine a bustling brewery, the smell of yeast and malt filling the air as wort is transformed into beer. This transformative process, known as fermentation, is where the alchemy of brewing truly happens. Just as ancient alchemists sought to turn base elements into gold, brewers seamlessly blend scientific knowledge with creative artistry to convert simple sugars into alcohol and the effervescent breath of carbon dioxide, creating the liquid gold we call beer.

In brewing, an alchemist is a brewer who masterfully manipulates chemical processes and techniques to produce unique and flavorful beverages. This term highlights the brewer's skill in transforming basic ingredients—water, malt, hops, and yeast—into beer, much like historical alchemists who sought to turn base materials into gold. But unlike those alchemists, brewers are truly successful in their transformative endeavors.

Understanding fermentation and the role of yeast is key to mastering brewing, as it significantly impacts the flavor, aroma, and overall quality of the final product. By exploring the intricacies of fermentation and yeast management through the lenses of rationalism, empiricism, and the scientific method, brewers can achieve greater control and consistency in their brewing practices.

Ancient Origins

Fermentation has a rich history dating back thousands of years. Ancient civilizations like the Sumerians and Egyptians practiced fermentation, though they lacked a scientific understanding of the process. They noticed that when certain grains were left in water, they would ferment and produce a drinkable, intoxicating beverage. This empirical observation was the foundation of early brewing, allowing these ancient cultures to produce and enjoy alcoholic drinks long before the biochemical mechanisms were understood.

They observed that natural processes transformed their ingredients into something new and delightful. Without knowledge of microorganisms, they attributed this transformation to mystical forces or the favor of deities. These early brewers laid the groundwork for fermentation techniques that would be refined over millennia.

Medieval Europe

During the medieval period, brewing techniques became more refined. Monastic breweries in Europe played a significant role in this evolution. The monks meticulously documented their brewing methods, leading to improved fermentation practices. They practiced careful record-keeping, detailing every step in the brewing process to track variables, identify patterns, and improve techniques.

These early advancements were largely empirical, with brewers learning through trial and error which conditions produced the best beer. They noticed that certain temperatures, ingredients, and methods yielded better results. The detailed records kept by the monks provided a foundation for more systematic approaches to brewing, combining empirical knowledge with rational insights.

Modern Era

The discovery of yeast by scientists like Louis Pasteur in the 19th century marked a turning point, transforming brewing from an art into a science. Pasteur's work on fermentation revealed the critical role of yeast in the brewing process and laid the groundwork for modern microbiology.

Understanding that yeast was responsible for fermentation allowed brewers to control the process more precisely. They began to manage yeast strains, ensuring purity and consistency. This understanding allowed brewers to control fermentation more precisely, improving the consistency and quality of their beer. Modern fermentation practices are deeply rooted in the scientific discoveries of this era, enabling brewers to achieve high levels of precision and consistency.

Rationalism—Deductive Reasoning

Biochemical Principles

Rationalism has been crucial in understanding the biochemical principles that govern fermentation. Through deductive reasoning, brewers have applied established theoretical frameworks to study yeast metabolism, temperature control, and fermentation kinetics. This rational, top-down deductive approach requires a systematic understanding of how various factors influence fermentation outcomes.

Yeast Metabolism: Yeast cells convert sugars into alcohol and carbon dioxide through metabolic pathways such as glycolysis and the citric acid cycle.

- Glycolysis is a metabolic pathway that converts glucose into pyruvate, yielding ATP and NADH. In brewing, glycolysis is essential as it leads to ethanol and carbon dioxide production during fermentation.
- The Citric Acid Cycle, also known as the Krebs cycle or TCA cycle, is a series of enzymatic reactions in yeast and other organisms crucial for cellular respiration. In brewing, it is the metabolic pathway where yeast converts sugars into energy, producing compounds that influence beer's flavors and aromas. This cycle occurs in the mitochondria of yeast cells and is essential for efficient energy production and yeast health during fermentation.

Understanding these pathways allows brewers to manipulate conditions to favor desired outcomes.

Temperature Control

The application of thermodynamic principles has guided the development of optimal temperature ranges for different yeast strains, ensuring efficient sugar conversion and desirable flavor profiles.

- Ale yeasts (top-fermenting) typically ferment best between 60-72°F (15-22°C).
- Lager yeasts (bottom-fermenting) require cooler temperatures, usually between 48-55°F (9-13°C).

By controlling fermentation temperatures within these optimal ranges, brewers can ensure consistent and high-quality fermentation. Temperature affects yeast activity and the production of flavor compounds such as esters and phenols.

- Esters are aromatic compounds produced during fermentation from the reaction between organic acids and alcohols. They contribute fruity and floral notes, such as banana, apple, and pear. The formation of esters is influenced by yeast strain, fermentation temperature, and wort composition.
- Phenols are chemical compounds produced during fermentation that impart flavors and aromas like spicy, smoky, clove, medicinal, and band-aid. Influenced by yeast strain, fermentation conditions, and raw materials, phenols are desirable in some beer styles (e.g., Belgian ales) but can be off-flavors in others.

By understanding and controlling these variables, brewers can influence the flavor profile of the beer.

Empiricism—Inductive Reasoning

Refining Practices

Empiricism, through careful observation and experimentation, has played a significant role in refining fermentation practices. Brewers have used inductive reasoning to gather data from countless fermentation batches, identifying patterns and optimizing processes. This hands-on approach has led to significant improvements in brewing techniques.

Observational Learning

Historical records show that brewers continually adjusted their fermentation techniques based on observations. They noted that certain yeast strains produced better flavors under specific conditions, leading to empirical adjustments that improved the fermentation process.

For example, they observed that higher fermentation temperatures led to increased ester production, resulting in fruity flavors. Conversely, lower temperatures reduced ester formation, producing a cleaner-tasting beer. These observations were crucial for developing effective fermentation strategies and enhancing the overall quality of beer.

Experimental Results

Empirical data from experiments conducted by brewers have driven many advancements. For example, experiments on pitching rates revealed how the amount of yeast added to the wort affects fermentation speed and flavor development.

- Pitching is the act of adding yeast to the wort to start fermentation. Proper pitching rates are essential for healthy and efficient fermentation.
- Under-pitching (adding too little yeast) can result in stressed yeast cells, leading to slow fermentation and the production of undesirable off-flavors such as diacetyl (buttery flavor) and acetaldehyde (green apple flavor).

- Diacetyl is a buttery-flavored compound produced by yeast during fermentation. A diacetyl rest allows the yeast to reabsorb and reduce this compound before fermentation is complete. This rest typically involves raising the temperature slightly at the end of primary fermentation.
- Acetaldehyde is a compound that can produce a green apple or raw pumpkin flavor in beer, typically resulting from incomplete fermentation.

- Over-pitching (adding too much yeast) can suppress the development of complex esters and phenols, resulting in a bland beer.

These findings have been incorporated into modern fermentation practices, demonstrating the importance of inductive reasoning in brewing. Experimentation has provided valuable insights that have shaped modern fermentation techniques.

The Scientific Method—Advancing Fermentation Techniques

Structured Approach

The scientific method provides a structured approach to advancing fermentation techniques. By applying observation, hypothesis formation, experimentation to verify or falsify the hypothesis, data analysis with assessment, replicability, and peer review, brewers have made significant strides in understanding and optimizing fermentation. This methodical approach ensures that brewing practices are based on solid evidence and rigorous testing.

Case Study: Temperature-Controlled Fermentation Chambers

One notable example is the development of temperature-controlled fermentation chambers. Brewers hypothesized that precise temperature control would improve fermentation consistency and beer quality.

- Experimentation: Brewers conducted trials using temperature-controlled environments versus ambient conditions, measuring fermentation rates, flavor profiles, and yeast health.
- Data Analysis: Results showed that temperature control led to more consistent fermentations, reduced off-flavors, and improved overall beer quality.
- Replicability and Peer Review: These findings were shared within the brewing community, leading to widespread adoption and further refinement of temperature control techniques.

This case study illustrates how the scientific method can lead to significant improvements in brewing practices.

The Fermentation Process

Yeast and Metabolic Pathways

Fermentation involves complex biochemical reactions that convert wort into beer. Yeast metabolizes fermentable sugars, producing alcohol and carbon dioxide as byproducts. This process is divided into three main stages:

1. Lag Phase
2. Primary Fermentation
3. Secondary Fermentation

Understanding these stages is essential for managing the fermentation process effectively.

Lag Phase and Primary Fermentation

- Lag Phase: The initial stage of fermentation where yeast acclimates to the environment, absorbs oxygen and nutrients, and prepares for cell division. This phase typically lasts:
    - 6-12 hours for ales
    - 12-18 hours for lagers

- Primary Fermentation: The stage of vigorous fermentation where the majority of sugar conversion occurs, producing alcohol and carbon dioxide.
    - Lasts 3-5 days for ales
    - Lasts 6-8 days for lagers
    - Indicated by foaming at the high krausen stage.
        - Krausen is the foamy head of yeast and proteins that forms on top of the wort during vigorous fermentation, indicating active yeast activity.

During primary fermentation, yeast activity is at its peak. Top-fermenting yeasts (ales) may produce a thick layer of krausen, which can be skimmed daily to remove excess yeast and proteins.

Temperature Control:

- Ale fermentation temperature: 65-68°F (18-20°C)
- Lager fermentation temperature: 46-50°F (8-10°C)

Secondary Fermentation

- Secondary Fermentation: A slower stage of fermentation after primary fermentation, where remaining complex sugars are converted, and the beer matures and clarifies.
    - For ales:
        - Indicated by the disappearance of foam and slower fermentation as most sugars are consumed.
        - Temperature remains the same as primary fermentation.
        - Lasts 1-3 days.
    - For lagers:
        - Secondary fermentation is gradual, with no abrupt drop-off in fermentation.
        - Specific gravity measurements are the best way to determine if primary fermentation is over.

- Starting about 24 hours after primary fermentation ends, the beer is cooled at a rate of 2-4°F/day until it reaches 32-34°F (0-1°C).
- Fermentation continues for 7-30 days.
- To minimize diacetyl, a diacetyl rest can be conducted immediately after primary fermentation, raising the temperature from 48°F to 60-65°F (15-18°C) for 24-48 hours.

Racking:

- Racking is the procedure of transferring beer from the primary fermentation vessel into another container for secondary fermentation. This process involves the risk of contamination but helps in conditioning and clarifying the beer.

Clarification Rest and Lagering

During maturation, the yeast reabsorbs and metabolizes by-products like acetaldehyde and diacetyl, and yeast cells flocculate and drop out, resulting in a clear beer.

- Flocculation: The clumping together and settling out of yeast cells after fermentation, affecting the clarity and yeast character of the beer.

Fining:

- Fining is a clarifying process that adds settling agents (like gelatin or isinglass) to precipitate colloidal matter, enhancing beer clarity.

Ales:

- Ales of normal gravity usually do not require much maturing.
- After secondary fermentation, “crash cool” the ale to 32-34°F (0-1°C) for 7-10 days to assist yeast dropout.
- Bottle Conditioning: Adding a small amount of fermentable sugar before bottling to allow yeast to produce natural

carbonation. Proper storage and handling are essential to avoid autolysis and oxidation.

- Autolysis: The self-digestion of yeast cells that occurs when they die and break down, potentially causing off-flavors such as a meaty or rubbery taste.
- Oxidation: A chemical reaction involving oxygen that can lead to off-flavors in beer, such as stale or cardboard-like tastes.

Yeast Management

Effective yeast management is vital for successful fermentation. Practices include proper pitching rates, aeration, temperature control, and sanitation.

Pitching Rates

Proper pitching rates ensure an adequate amount of healthy yeast for fermentation.

- Under-pitching can lead to slow fermentation and the development of undesirable flavors.
- Over-pitching can result in a lack of yeast character in the beer.

Aeration

Yeast requires oxygen during the initial growth phase to build cell walls and reproduce.

- Aerate the wort before pitching yeast to provide necessary oxygen.
- Avoid aeration after fermentation starts, as it can cause oxidation and spoil the beer.

Closed Fermentation

- Closed fermentation refers to fermenting beer in a sealed vessel that prevents exposure to outside air. This method helps maintain a controlled environment, protecting the beer from contamination by wild yeasts, bacteria, and oxygen.

- The fermenter should be at least one-third oversized to accommodate the foaming action during vigorous fermentation.
- Closed fermentation is especially crucial for brewing lagers and other styles that require precise control over fermentation conditions.

Yeast Starters and Washing

- Yeast Starter: A yeast starter is a small batch of wort used to propagate yeast cells before pitching them into the main batch of beer. This process increases the yeast cell count and vitality, leading to a more vigorous and complete fermentation.
  - Mix water, malt extract, and yeast.
  - Allow the yeast to multiply and become more active.
  - Especially useful for high-gravity beers, lagers, or when using older yeast.
- Yeast Washing: Yeast washing is a process used to clean and purify harvested yeast for reuse in future batches.
  - Separates yeast from trub (protein, hop residue, and other debris) and unwanted microorganisms.
  - Involves adding sterile water or a dilute acid solution to a yeast slurry, allowing the mixture to settle, and then decanting the clean yeast from the top layer.
  - Helps maintain yeast vitality and health, ensuring consistent fermentation performance and preventing off-flavors.

Challenges

Consistent Fermentation Temperatures

Maintaining consistent fermentation temperatures within the yeast strain's optimal range is critical for successful fermentation.

- Use temperature-controlled fermentation chambers or water baths to regulate the environment.

- Prevent temperature fluctuations that can stress yeast and lead to off-flavors.
- Regularly monitor gravity with a hydrometer or refractometer to track fermentation progress and determine when fermentation is complete.

Managing Fermentation Stages

- Stuck Fermentation: Occurs when yeast ceases activity before all fermentable sugars are consumed.
    - Causes: Poor yeast health, insufficient nutrients, temperature fluctuations.
    - Solutions:
        - Increase temperature slightly to re-activate yeast.
        - Add yeast nutrients.
        - Pitch additional active yeast.
- Off-Flavors: Can result from stressed yeast, infection, or poor ingredient quality.
    - Common off-flavors include diacetyl, acetaldehyde, and fusel alcohols.
        - Fusel Alcohols are higher alcohols produced during fermentation, especially at higher temperatures. They contribute strong, solvent-like flavors and aromas, described as hot, spicy, or alcoholic.
    - Solutions:
        - Proper yeast management.
        - Sanitation.
        - Ingredient selection.
        - Control fermentation temperature.
- Contamination: Caused by unwanted microorganisms such as wild yeast or bacteria.

- Rigorous sanitation practices and monitoring can prevent contamination.

Yeast Health and Viability

- Ensure yeast is fresh and viable before pitching.
- Use yeast nutrients to provide essential minerals and vitamins.
- Monitor yeast performance and adjust practices as needed.

Ongoing Research and Future Directions

Discussing ongoing research and potential future directions for improving fermentation techniques keeps brewers informed about the latest advancements in the field.

Genetically Modified Yeast

Advances in biotechnology have led to the creation of yeast strains with specific genetic modifications that enhance their performance in fermentation.

- Yeast strains engineered to:
    - Produce lower levels of unwanted byproducts.
    - Ferment at higher temperatures without producing off-flavors.
    - Enhance flavor profiles with novel compounds.

Advanced Monitoring Tools

The use of modern technology, such as real-time monitoring of fermentation parameters (e.g., temperature, pH, dissolved oxygen), can provide brewers with immediate feedback.

- Allows for prompt adjustments.
- Ensures optimal fermentation conditions.
- Improves consistency and quality.

Alternative Fermentation Methods

Research into alternative fermentation methods, such as mixed fermentation with multiple yeast strains and bacteria, can create unique and complex flavor profiles.

- Mixed Fermentation: The use of multiple yeast strains and/or bacteria to ferment beer, often resulting in complex and unique flavor profiles.
- Expands the diversity of beer styles.
- Offers creative opportunities for brewers.

Conclusion

Fermentation is a transformative process where yeast converts wort into beer, significantly impacting flavor, aroma, and overall quality. Mastery of fermentation is essential for brewing, involving historical insights, rationalism, empiricism, and the scientific method.

- Ancient civilizations like the Sumerians and Egyptians observed fermentation empirically.
- Medieval monastic breweries refined practices through meticulous documentation.
- The 19th-century discovery of yeast by Louis Pasteur allowed for precise control, enhancing the process.

Rationalism aids in understanding biochemical principles like yeast metabolism and temperature control. Empiricism refines practices through observation and experimentation. The scientific method advances techniques through structured approaches, such as temperature-controlled chambers.

By optimizing fermentation stages and applying meticulous yeast management, brewers can achieve consistent, high-quality results. As brewing technology and knowledge advance, ongoing research and best practices will further enhance the craft, fostering continuous improvement and innovation in the brewing industry.

> *To ferment is to entrust life to the unseen—where intention meets transformation, and the brewer's truest virtue is patience in becoming.*

**

Review Questions

True/False Questions

1. True or False: Louis Pasteur discovered the role of yeast in fermentation, transforming brewing from an art into a science.

2. True or False: Empiricism has no significant role in refining fermentation practices.

3. True or False: Secondary fermentation occurs after the primary fermentation and is crucial for conditioning and clarifying the beer.

4. True or False: Maintaining consistent fermentation temperatures is critical for achieving high-quality fermentation.

5. True or False: Under-pitching yeast can lead to the production of undesirable off-flavors such as diacetyl and acetaldehyde.

Multiple Choice Questions

6. Which stage of fermentation involves yeast acclimatizing to the new environment and preparing for cell division?
A) Secondary fermentation
B) Primary fermentation
C) Lag phase
D) Clarification rest

7. What is a common temperature range for fermenting ales?
A) 48-55°F (9-13°C)
B) 32-34°F (0-1°C)
C) 60-72°F (15-22°C)
D) 40-45°F (4-7°C)

Brewer Vignettes

8. Brewer Vignette 1: Imagine you are a brewer aiming to prevent stuck fermentation in your beer. Describe the steps you would take to ensure a smooth fermentation process.
A) Increase the fermentation temperature and pitch additional active yeast if necessary.
B) Avoid aerating the wort before pitching yeast.
C) Use random pitching rates without monitoring.
D) Skip secondary fermentation altogether.

9. Brewer Vignette 2: As a brewer, you want to enhance the flavor complexity of your beer. Explain how you would manage the fermentation process to achieve this.
A) Ferment at a very low temperature to slow down yeast activity.
B) Experiment with different yeast strains and manage pitching rates, fermentation temperatures, and aeration to optimize flavor development.
C) Avoid using any hops during fermentation.
D) Skip the secondary fermentation stage.

10. Brewer Vignette 3: You are noticing off-flavors such as diacetyl in your beer. Describe the methods you would use to address and prevent this issue.
A) Increase the fermentation temperature abruptly.
B) Conduct a diacetyl rest by raising the temperature towards the end of primary fermentation, then lower it gradually for secondary fermentation.
C) Skip aerating the wort before pitching yeast.
D) Use under-pitching rates to slow down fermentation.

**

Correct Answers

True/False Questions

1. True
2. False (Empiricism has played a significant role in refining fermentation practices through observation and experimentation.)
3. True
4. True
5. True

Multiple Choice Questions

6. C) Lag phase
7. C) 60-72°F (15-22°C)

Brewer Vignettes

8. A) Increase the fermentation temperature and pitch additional active yeast if necessary.
9. B) Experiment with different yeast strains and manage pitching rates, fermentation temperatures, and aeration to optimize flavor development.
10. B) Conduct a diacetyl rest by raising the temperature towards the end of primary fermentation, then lower it gradually for secondary fermentation.

**

Beyond The Chapter

- Bamforth, C. W. (2003). Beer: Tap into the Art and Science of Brewing. Oxford University Press.
  - *Discusses the fermentation process and its critical role in beer production.*
- Palmer, J. (2006). How to Brew: Everything You Need To Know To Brew Beer Right The First Time. Brewers Publications.
  - *Provides practical advice on managing fermentation for optimal beer quality.*
- White, C., & Zainasheff, J. (2010). Yeast: The Practical Guide to Beer Fermentation. Brewers Publications.

    - *A comprehensive guide on yeast management and fermentation techniques.*
- Noonan, G. (1996). Brewing Lager Beer: The Most Comprehensive Book for Home - And Microbrewers. Brewers Publications.
    - *Focuses on fermentation methods for producing high-quality lagers.*
- Kunze, W. (2014). Technology Brewing & Malting. VLB Berlin.
    - *Covers the technical aspects of fermentation and modern equipment used.*
- Pasteur, L. (1876). Studies on Fermentation: The Diseases of Beer. Macmillan and Co.
    - *Presents Pasteur's pioneering research on fermentation and its applications in brewing.*
- Hansen, E. C. (1883). Practical Studies in Fermentation. Carlsberg Laboratory Publications.
    - *Explores early scientific studies on yeast and fermentation processes.*
- Krogerus, K., & Gibson, B. (2013). "Yeast Genetics: The Biochemistry and Molecular Biology of Wine Yeasts". Yeast.
    - *Discusses advanced topics in yeast genetics and their implications for brewing.*

Weblinks

- Understanding Fermentation - Brew Your Own: https://byo.com/?s=Fermentation&post_type=article&orderby=relevance
- Yeast and Fermentation - Craft Beer & Brewing Magazine: https://beerandbrewing.com/search/?search=fermentation
- Fermentation Basics - Homebrew Academy: https://homebrewacademy.com/?s=fermentation
- The Science of Fermentation - American Homebrewers Association: https://www.homebrewersassociation.org/?s=Fermentation

These references provide deeper insights into fermentation and yeast management techniques, offering valuable resources for both homebrewers and professionals.

***

Chapter 8: The Art and Science of Dry Hopping—Crafting the Essence of Modern Beer

> *Dry hopping is where the science of hops meets the artistry of aroma, crafting the essence of modern beer.*

Imagine the burst of citrusy, floral, and piney aromas that hits you when you crack open a fresh IPA. These floral aromas, reminiscent of blooming flowers, are common in hop varieties like Cascade and Saaz. The piney notes, evocative of pine needles or resin, are characteristic of hops such as Chinook, Simcoe, and Columbus. This explosion of scent and flavor is the hallmark of dry hopping, a transformative brewing technique that has revolutionized modern beer, especially hop-forward styles like IPAs.

Unlike traditional hop additions during the boil, which primarily contribute bitterness through the isomerization of alpha acids, dry hopping focuses on enhancing the aromatic profile of the beer. By adding hops to the beer after the boil and primary fermentation, brewers can enhance aroma and flavor without adding significant bitterness. Understanding the principles of dry hopping and how to apply them allows brewers to create beers with vibrant hop aromas and flavors that captivate the senses. This technique has become essential in crafting beers that stand out in today's competitive market.

Early Uses of Hops

The practice of dry hopping has evolved significantly over the centuries. In the early days of brewing, hops were primarily used for their preservative qualities and bitterness. Ancient brewers discovered that adding hops at different times during brewing could change the taste and smell of the beer. Through trial and error, they realized that late-stage hopping profoundly impacted aroma and flavor.

The Rise of Dry Hopping

The modern craft beer movement, which gained momentum in the late 20th century, embraced dry hopping as a way to differentiate their beers and appeal to a growing market of hop enthusiasts. Pioneers in the craft beer industry, such as Ken Grossman of Sierra Nevada Brewing Company, played a significant role in popularizing dry hopping. This historical journey highlights the continuous interplay between rationalism and empiricism in refining brewing techniques. Dry hopping has since become a standard practice in many breweries, contributing to the distinct and desirable characteristics of modern craft beers.

Rationalism—Deductive Reasoning

Chemical Principles of Dry Hopping

Rationalism has played a crucial role in understanding the chemical principles underlying dry hopping. Through deductive reasoning, brewers have applied established theoretical frameworks to study hop compounds, such as essential oils and alpha acids. This scientific approach has allowed brewers to optimize their processes and achieve consistent results.

For example, understanding the volatility of hop oils has guided brewers in determining the optimal timing for dry hopping to maximize aromatic impact. Volatility refers to the tendency of a substance to vaporize. In hops, the volatile oils responsible for aroma and flavor are sensitive to heat and can be easily lost during boiling.

Essential oils such as myrcene, which imparts citrusy, piney, and herbal aromas; humulene, known for earthy and woody notes; and caryophyllene, providing spicy and woody characteristics, are key components of hop aroma. The high volatility of compounds like myrcene and caryophyllene makes dry hopping an effective method for preserving these delicate scents. Humulene, being more stable, contributes to aroma both during boiling and dry hopping.

Guided Techniques

Rationalism provides a top-down approach, using theoretical knowledge to inform practical applications and improve brewing outcomes. Brewers have determined that adding hops after primary fermentation allows for better retention of aromatic compounds, as the absence of active fermentation prevents the loss of these volatile oils through CO_2 scrubbing. This understanding has led to more precise and effective dry hopping techniques, enhancing the overall quality of the beer.

Empiricism—Inductive Reasoning

Refining Dry Hopping Practices

Empiricism, through careful observation and experimentation, has significantly refined dry hopping practices. Brewers have used inductive reasoning to gather data from countless dry hopping experiments, identifying patterns and optimizing processes. This iterative approach has been essential in developing effective dry hopping strategies.

Observational Learning

Historical records show that brewers continually adjusted their dry hopping techniques based on observations. They noted that certain hop varieties and contact times—the duration hops remain in contact with the beer during dry hopping—produced better aromas and flavors. Longer contact times can intensify aromas but may introduce grassy or vegetal flavors. These observations led to empirical adjustments that improved the process, helping brewers find the optimal balance between intensity and undesirable off-flavors.

Experimental Results

Empirical data from experiments conducted by brewers have driven many advancements. For example, experiments on the timing of dry hopping revealed that adding hops after primary fermentation but before conditioning results in the most intense hop aroma and flavor. Continuous experimentation has allowed brewers to refine their techniques and produce beers with exceptional aromatic profiles, demonstrating the importance of inductive reasoning in brewing.

The Scientific Method—Advancing Dry Hopping Techniques

Structured Approach

The scientific method provides a structured approach to advancing dry hopping techniques. By applying observation, hypothesis formation, experimentation to verify or falsify the hypothesis, data analysis with assessment, replicability, and peer review, brewers have made significant strides in understanding and optimizing dry hopping. This systematic approach ensures that improvements are based on solid evidence and reproducible results.

Case Study: Hop Sensory Analysis

One notable example is the development of hop sensory analysis techniques. Brewers hypothesized that different hop varieties would contribute distinct aromatic profiles. Experiments confirmed this hypothesis, leading to the creation of hop flavor wheels and sensory analysis methods—the practice of evaluating beer characteristics, such as aroma, flavor, and mouthfeel, through sensory methods to refine brewing techniques. Data analysis and peer review within the brewing community have further refined these techniques, enhancing the ability to craft hop-forward beers. This case study illustrates the power of the scientific method in driving innovation and improving brewing practices.

Purpose of Dry Hopping

Enhancing Aroma and Flavor

The primary purpose of dry hopping is to enhance the hop aroma and flavor in the finished beer. Unlike traditional hop additions during the boil, which primarily contribute bitterness due to the isomerization of alpha acids, dry hopping emphasizes the aromatic oils of hops. These oils are responsible for the floral, citrusy, piney, and fruity notes that define many modern beers.

By adding hops after the primary fermentation, brewers can preserve these volatile compounds, resulting in a more vibrant and aromatic beer. This technique has become essential in creating the bold and complex hop profiles that beer enthusiasts crave.

Timing of Dry Hopping

Optimal Timing

Timing is crucial for dry hopping. It is typically performed after primary fermentation because the active bubbling of carbon dioxide can drive off volatile aromatic compounds, diminishing their impact. By waiting until fermentation has slowed or stopped, brewers can maximize the retention of these desirable compounds.

Duration of Dry Hopping

The duration of dry hopping can vary significantly, ranging from a few days to a couple of weeks. The longer the hops remain in contact with the beer, the more intense the hop aroma and flavor will be. However, extended contact times can sometimes lead to grassy or vegetal flavors due to the extraction of less desirable compounds. Finding the right balance is key. Brewers must carefully monitor the dry hopping process to achieve the desired aromatic profile without introducing undesirable flavors.

Hop Utilization

Concept of Hop Utilization

Understanding hop utilization is essential for brewers aiming to craft unique and flavorful beers. Hop utilization refers to how effectively the hops impart their desired qualities—bitterness, flavor, and aroma—into the beer. Factors such as hop variety, form, and addition timing all influence hop utilization. By understanding these factors, brewers can optimize their hop usage and create the desired flavor profile.

Factors Affecting Utilization

Higher utilization rates are achieved during the boil, where alpha acids are isomerized, adding bitterness. In contrast, lower temperatures during dry hopping preserve aromatic compounds without adding significant bitterness. This understanding allows brewers to strategically use hops to achieve the desired balance of bitterness and aroma in their beers.

Hop Varieties

Popular Hop Varieties

Different hop varieties impart unique flavors and aromas, making the selection of hops a crucial decision in the brewing process. For instance:

- Cascade hops are known for their citrus and floral notes.
- Citra hops offer intense tropical fruit flavors.
- Chinook, Simcoe, and Columbus hops provide piney and resinous characteristics.

Each hop variety contributes distinct characteristics, allowing brewers to create diverse and complex beer profiles.

Experimentation with Hops

Experimenting with various hop varieties can help brewers create diverse and complex beer profiles. Blending different hops can enhance the overall aroma and flavor, providing a more balanced and layered sensory experience. By trying new hop combinations and techniques, brewers can continuously innovate and improve their beers.

Hop Forms

Different Forms of Hops

Hops are available in several forms, each with its own benefits and drawbacks.

- Whole Hops: These are the dried hop cones as they are harvested. Whole hops are less processed and can provide a more natural hop character but can be less efficient in terms of storage and utilization. They are often preferred by traditionalists who value their natural properties.
- Pellet Hops: Pellet hops are made by grinding whole hops into a powder and then pressing them into small pellets. They offer better stability and higher utilization rates compared to whole hops. Pellet hops are easier to store, measure, and use, making them widely popular for their convenience and effectiveness.
- Hop Extracts: Concentrated hop oils and resins used for precise control over bitterness and aroma. They offer consistency and ease of use but may lack the complexity and depth of flavor provided by whole or pellet hops. Hop extracts are valuable for delivering specific hop characteristics with precision, especially in large-scale commercial brewing.

Challenges

Hop Storage

Proper storage is critical to maintaining hop freshness and potency. Hops should be stored in a cool, dark place, preferably vacuum-sealed, to protect them from oxygen and light, both of which can degrade hop quality. Refrigeration or freezing is recommended for long-term storage. Proper storage practices ensure that hops retain their aromatic and flavor qualities, contributing to the overall quality of the beer.

Avoiding Oxidation

Minimize oxygen exposure during dry hopping to preserve hop character and prevent oxidation. Use purging techniques—

removing oxygen from the vessel by filling it with CO_2 before adding the hops—to displace any oxygen. Purging helps preserve beer freshness and prevent oxidation, which can dull hop flavors and introduce off-flavors.

Additionally, consider using hop bags or hop cages—devices used to contain hops during dry hopping—to facilitate easier removal and minimize the risk of oxygen ingress and contamination. These practices help maintain the quality and freshness of the hops during the dry hopping process.

Common Challenges

Dry hopping can present several challenges, including hop creep, hop burn, and grassy or vegetal flavors.

- Hop Creep: Refers to the unintended continued fermentation caused by enzymes in hops during dry hopping. This can lead to changes in carbonation levels and potential off-flavors. To mitigate hop creep, brewers can pasteurize the beer after dry hopping or choose hop varieties with lower enzymatic activity.
- Hop Burn: A harsh, astringent sensation often associated with over-dry hopping. It can be mitigated by adjusting hop quantities and contact time to prevent the extraction of harsh compounds.

Addressing these issues requires a combination of empirical observation and theoretical understanding. Regular monitoring and adjustment of dry hopping practices based on sensory analysis and chemical testing can help brewers overcome these challenges and consistently produce high-quality beers. By staying vigilant and proactive, brewers can ensure that their dry hopping techniques yield the best possible results.

Conclusion

Dry hopping is a nuanced and impactful technique that enhances the aromatic profile of beer. By understanding the interplay between rationalism, empiricism, and the scientific method, brewers can refine their dry hopping practices to achieve desired results. This chapter demonstrates how traditional knowledge and modern science combine to elevate the craft of brewing. Mastery of dry hopping allows brewers to create beers with captivating aromas and flavors that delight the senses.

> *Dry hopping reminds us that pleasure is not frivolous—it is a form of care. To fill the air with fleeting beauty is to say: this, too, matters.*

**

Review Questions

True/False Questions

1. True or False: The primary purpose of dry hopping is to enhance the hop aroma and flavor in the finished beer.

2. True or False: Dry hopping is typically performed during the primary fermentation stage.

3. True or False: Pellet hops are the most common form used by homebrewers due to their ease of storage and utilization.

4. True or False: Empirical data and experimentation have played a significant role in refining dry hopping practices.

5. True or False: Whole hops are the least processed form of hops and can provide a more natural hop character.

Multiple Choice Questions

6. Which of the following hop varieties is known for its intense tropical fruit flavors?
A) Cascade
B) Centennial
C) Amarillo
D) Citra

7. What is the optimal timing for dry hopping to maximize aromatic impact?
A) Before primary fermentation
B) During primary fermentation
C) After primary fermentation but before conditioning
D) During the boil

Brewer Vignettes

8. Brewer Vignette 1: You want to avoid grassy or vegetal flavors in your dry-hopped beer. What is the best approach?
A) Use extended contact times of several weeks.
B) Add hops during the primary fermentation.
C) Monitor dry hopping duration carefully and avoid extended contact times.
D) Use only whole hops for dry hopping.

9. Brewer Vignette 2: As a brewer aiming to prevent oxidation during dry hopping, what technique should you use?
A) Add hops directly without any precautions.
B) Flush the dry hopping vessel with CO_2 before adding hops.
C) Store hops at room temperature before use.
D) Use only hop extracts.

10. Brewer Vignette 3: You are experimenting with hop combinations to create a complex beer profile. Which approach would be most effective?
A) Stick to a single hop variety for simplicity.
B) Blend different hops and adjust the timing of additions to layer aromas and flavors.
C) Avoid dry hopping and focus on bittering hops.
D) Use only late hop additions during the boil.

**

Correct Answers
True/False Questions
1. True
2. False (Dry hopping is typically performed after the primary fermentation stage.)
3. True
4. True
5. True

Multiple Choice Questions
6. D) Citra
7. C) After primary fermentation but before conditioning

Brewer Vignettes
8. C) Monitor dry hopping duration carefully and avoid extended contact times.
9. B) Flush the dry hopping vessel with CO_2 before adding hops.
10. B) Blend different hops and adjust the timing of additions to layer aromas and flavors.

**

Beyond The Chapter

- Bamforth, C. W. (2003). Beer: Tap into the Art and Science of Brewing. Oxford University Press.
    - *Explores the use of hops in brewing and the impact of dry hopping on beer flavor.*
- Palmer, J. (2006). How to Brew: Everything You Need To Know To Brew Beer Right The First Time. Brewers Publications.
    - *Provides practical guidance on hop utilization and dry hopping techniques.*
- Hieronymus, S. (2012). For the Love of Hops: The Practical Guide to Aroma, Bitterness, and the Culture of Hops. Brewers Publications.
    - *A detailed guide on the use of hops in brewing, including dry hopping methods.*

- Noonan, G. (1996). Brewing Lager Beer: The Most Comprehensive Book for Home - And Microbrewers. Brewers Publications.
    - *Discusses hop utilization techniques specific to lager production.*
- Daniels, R. (1996). Designing Great Beers: The Ultimate Guide to Brewing Classic Beer Styles. Brewers Publications.
    - *Examines the role of hops in different beer styles and their impact on flavor and aroma.*

Weblinks

- Dry Hopping Techniques - Craft Beer & Brewing Magazine: https://beerandbrewing.com/search/?search=Dry+Hopping
- The Art of Dry Hopping - Brew Your Own: https://byo.com/?s=Dry+Hopping
- Dry Hopping Best Practices - American Homebrewers Association: https://www.homebrewersassociation.org/?s=Dry+hopping
- The Science of Dry Hopping - Homebrew Academy: https://homebrewacademy.com/?s=Dry+Hopping
- Hop Varieties and Utilization - Brew Cabin: https://www.brewcabin.com
- Dry Hopping Practices - Sierra Nevada Brewing Co.: https://sierranevada.com/knowledge-base?snbc_search%5Bquery%5D=hops

These references provide deeper insights into dry hopping techniques, offering valuable resources for both homebrewers and professionals.

***

Chapter 9: The Final Touch—Bottling and Kegging for the Perfect Pour

Bottling and kegging are the final acts of precision, preserving the integrity and flavor of every brew for the perfect pour.

Imagine the clink of glass bottles and the hiss of keg taps as beer is prepared for its final journey to the consumer. Bottling and kegging are crucial final steps in the brewing process that ensure beer is preserved, properly carbonated, and ready for consumption. These stages not only prepare the beer for storage and serving but also play a significant role in maintaining its quality, flavor, and freshness. By mastering the techniques of bottling and kegging, brewers can consistently deliver high-quality beer. This chapter explores these processes through the lenses of rationalism, empiricism, and the scientific method, demonstrating how each contributes to the art and science of brewing. The attention to detail and precision in these stages can make a significant difference in the final product.

Evolution of Bottling and Kegging Techniques

Early Storage Methods

The methods of bottling and kegging have undergone significant evolution over time. In the early days of brewing, beer was stored in wooden barrels and stoneware bottles. Wooden barrels, often made from oak, were used for fermenting and aging beer. These barrels imparted complex flavors such as vanilla, caramel, and spices due to interactions between the beer and the wood. They played a crucial role in early brewing techniques, offering natural insulation and slight oxygenation that influenced the aging process. Proper maintenance and sanitation were essential to prevent spoilage and contamination, ensuring the beer's quality and flavor.

Stoneware bottles were an early type of beer container made from ceramic materials before the widespread adoption of glass bottles. They offered limited control over carbonation and preservation. Stoneware bottles were durable but heavy and opaque, making it difficult to monitor the beer's clarity and carbonation levels.

Introduction of Glass Bottles

The invention of glass bottles in the 17th century marked a major advancement, allowing brewers to better preserve beer and control carbonation levels. Glass bottles were transparent, lightweight, and non-reactive, making them ideal for beer storage. This shift not only improved the quality of beer but also made it more accessible to a broader audience.

The development of the crown cap, a metal cap used to seal beer bottles, further enhanced the bottling process. Crown caps prevent contamination and maintain carbonation by creating an airtight seal. They are crimped onto the bottle using a capping tool, ensuring the beer remains fresh and carbonated.

Advancements in Kegging

The introduction of metal kegs in the 20th century revolutionized beer storage and transport. A keg is a cylindrical container used for storing and dispensing beer. Kegs are typically made of stainless steel and can be reused multiple times, offering a durable and efficient packaging solution. They protect the beer from light and oxygen, preserving its freshness and flavor over extended periods.

Scientific Innovations

Scientific advancements, such as the understanding of microbial contamination and the development of pasteurization techniques, further refined bottling and kegging methods. Innovations like CO_2 purging, which uses carbon dioxide to displace oxygen in bottles or kegs to prevent oxidation, have enhanced the reliability and consistency of beer packaging. Modern sanitation practices, including the use of effective sanitizers and strict hygiene protocols, ensure that beer remains uncontaminated throughout the packaging process.

These developments have allowed brewers to produce and distribute beer on a larger scale without compromising quality. The evolution of packaging materials and techniques reflects the continuous interplay between tradition and innovation in brewing.

Rationalism—Deductive Reasoning

Principles Behind Bottling and Kegging

Rationalism has been instrumental in developing the principles underlying bottling and kegging. Top-down deductive reasoning has allowed brewers to formulate theories on carbonation, sanitation, and storage. For instance, understanding the physics of carbonation has led to precise methods for achieving desired CO_2 levels in beer.

Theories on microbial contamination have informed stringent sanitization protocols. Sanitization—the practice of cleaning and disinfecting brewing equipment to eliminate unwanted microorganisms—ensures the beer remains uncontaminated throughout the packaging process. This theoretical approach has enabled brewers to anticipate and mitigate potential issues before they arise.

Theoretical Frameworks and Established Principles

Key theoretical frameworks have guided the development of modern bottling and kegging practices. The principles of gas laws, particularly those involving pressure and temperature, have been applied to design effective carbonation techniques. Understanding the biochemical stability of beer has led to innovations in storage and preservation methods.

For example, the concept of force carbonation—a method of carbonating beer by directly infusing CO_2 into a keg under pressure—allows for precise control over carbonation levels and quicker results compared to natural carbonation. These rational approaches ensure that each step in the bottling and kegging process is optimized for maintaining beer quality. By applying these principles, brewers can create a consistent and enjoyable drinking experience.

Empiricism—Inductive Reasoning

Refining Bottling and Kegging Techniques

Empiricism, based on careful observation and experimentation, has played a crucial role in refining bottling and kegging techniques. Brewers have gathered data from numerous packaging experiments, using inductive reasoning to identify patterns and optimize processes. Empirical evidence has informed best practices, ensuring that these techniques are both scientifically sound and practically effective. This iterative process of testing and refinement has led to significant improvements in the quality and efficiency of beer packaging.

Observations and Experiments

Historical records and modern experiments have shown how different bottling and kegging methods impact beer quality. For example, empirical studies have demonstrated the effects of various sanitizers on beer stability and flavor. Experimentation with different carbonation techniques has led to the development of reliable methods for achieving consistent CO_2 levels.

Brewers observed that certain sanitizers effectively eliminate microorganisms without leaving residual flavors. They also found that varying the amount and type of priming sugar—sugar added to beer before bottling to induce fermentation and natural carbonation—affects the carbonation level and mouthfeel of the final product. Common types of priming sugar include corn sugar (dextrose), cane sugar (sucrose), and dry malt extract (DME).

These observations and experiments have been crucial in validating theoretical models and improving practical applications.

By continuously observing and experimenting, brewers can adapt to new challenges and innovations in the field.

The Scientific Method—Advancing Bottling and Kegging

Structured Approach to Improvement

The scientific method provides a structured approach to enhancing bottling and kegging techniques. By following the steps of observation, hypothesis formation, experimentation to verify or falsify the hypothesis, data analysis with assessment, replicability, and peer review, brewers have made significant advances in understanding and optimizing these processes.

Case Studies

One notable example is the development of precise carbonation methods. Brewers hypothesized that force carbonation could provide more consistent results than natural carbonation. Experiments confirmed this hypothesis, leading to widespread adoption of force carbonation in kegging.

Another case study involves the effects of different sanitizers on beer quality. Brewers tested various sanitizing agents to determine their efficacy and impact on flavor. Data analysis and peer review within the brewing community have further refined these techniques. This systematic approach ensures that improvements are based on solid evidence and can be replicated by others in the brewing community.

Bottling

Sanitization

When bottling, sanitization is paramount to prevent contamination that could spoil the beer or introduce off-flavors. Thoroughly cleaning all equipment, including bottles, caps, siphons, and bottling wands, is essential. High-quality no-rinse sanitizers, such as Star San or iodophor, ensure all surfaces are free from harmful microorganisms. Even minimal contamination can ruin an entire batch, making sanitization one of the most critical steps in bottling.

To sanitize:

- Immerse bottles and caps in the sanitizer solution, ensuring every surface comes into contact with the solution.
- Allow the bottles to drain and dry on a sanitized bottle tree or rack.
- For siphons and tubing, fill them with the sanitizer and let them sit for the recommended contact time before draining.

This thorough approach helps maintain the purity and quality of the beer.

Priming Sugar

Before bottling, priming sugar is added to the beer so that the yeast can provide natural carbonation—a process known as natural carbonation. The sugar ferments inside the bottle, producing CO_2 that carbonates the beer.

Common types of priming sugar include:

- Corn sugar (dextrose)
- Cane sugar (sucrose)
- Dry malt extract (DME)

Each type of sugar will ferment slightly differently, so it's essential to use a priming sugar calculator to determine the correct amount. Factors to consider include:

- Beer style
- Desired carbonation level
- Volume of beer to be bottled

To add priming sugar:

1. Dissolve the measured amount of sugar in a small amount of boiling water to create a simple syrup. Boiling ensures the sugar is sterile.
2. Cool the solution slightly.

3. Gently stir it into the beer to ensure even distribution without introducing oxygen, which can cause oxidation and off-flavors.

This method ensures consistent carbonation across all bottles.

Bottle Conditioning

Bottle conditioning is the process where the beer undergoes fermentation in the bottle, creating natural carbonation. After bottling, the beer should be stored at room temperature, around 68-72°F (20-22°C), for 1-2 weeks. During this period, the yeast will consume the priming sugar, producing CO_2 and carbonating the beer.

Patience is key during bottle conditioning. Resist the urge to open a bottle too early, as premature opening can result in under-carbonated beer. After the initial conditioning period, move the bottles to a cool, dark place to allow any remaining yeast to settle and the flavors to mature. Storing bottles upright minimizes yeast sediment disturbance, resulting in a clearer pour. This careful conditioning process ensures that the beer develops its full flavor and carbonation profile.

Kegging

Carbonation Methods

There are two primary methods for carbonating beer in a keg:

1. Force Carbonation
    - Involves directly adding CO_2 to the kegged beer.
    - Connect the keg to a CO_2 tank and adjust the regulator to the desired pressure, typically around 12-14 psi for most beer styles.
    - Place the keg in a cold environment, such as a kegerator—a refrigerated unit designed to hold kegs and dispense beer, equipped with a tap and CO_2 system to maintain serving pressure and temperature.
    - Leave it under pressure for 24-48 hours.

- This method allows for precise control over carbonation levels and is much quicker than natural carbonation.

2. Natural Carbonation
   - Similar to bottle conditioning.
   - Add priming sugar to the beer before sealing the keg.
   - Allow the beer to carbonate over several weeks at room temperature.
   - Less common due to the availability and convenience of force carbonation.
   - Some brewers prefer it for the unique flavor profile it can impart.

Serving Pressure

Maintaining the appropriate serving pressure is crucial for a consistent pour and carbonation level. The serving pressure—the pressure at which beer is dispensed from a keg—is typically maintained between 8-12 psi but can vary based on:

- Beer style
- Keg temperature
- Desired carbonation

Monitor the pressure regularly and adjust as necessary:

- Serving beer at too high a pressure can result in overly foamy pours.
- Too low a pressure can cause flat, under-carbonated beer.
- Start with the recommended pressure for the beer style and make minor adjustments to achieve the perfect pour.

This careful management ensures that each glass of beer is as enjoyable as the brewer intended.

Challenges

Bottle Storage

- Store bottles upright to minimize yeast sediment disturbance.
- Protect bottles from light exposure to prevent skunking—a light-induced off-flavor.
- Maintain a consistent, cool temperature to preserve beer freshness.

Keg Maintenance

Regularly clean and sanitize kegs and serving lines to maintain beer quality:

- After each keg is emptied, disassemble all parts, including the dip tube, poppets, and posts.
- Clean thoroughly with a suitable cleaner.
- Rinse well and sanitize before reassembling.
- Clean serving lines with a line cleaner every few kegs to prevent buildup and off-flavors.

This routine maintenance is crucial for preserving the taste and integrity of the beer.

Avoiding Oxidation

When bottling or kegging, take care to minimize oxygen exposure:

- Use CO_2 to purge the headspace of bottles and kegs before sealing.
- Fill bottles and kegs gently to reduce splashing and aeration.
- Oxygen exposure can lead to oxidation, causing stale, cardboard-like flavors.

Proper handling techniques ensure that the beer retains its intended flavor profile throughout its shelf life.

Common Challenges

Bottling and kegging present several challenges, including:

- Contamination: Ensuring thorough sanitization of all equipment and maintaining clean working conditions are crucial for preventing contamination. Regularly inspect and replace any equipment showing signs of wear or damage.
- Under/Over-Carbonation: Properly calculating and evenly distributing priming sugar is essential for achieving the desired carbonation level. Regularly check and adjust serving pressure to ensure consistent carbonation in kegged beer.
- Oxidation: Minimizing oxygen exposure during bottling and kegging is crucial for preserving beer quality. Purging bottles and kegs with CO_2 and using oxygen-absorbing caps can help prevent oxidation.

Addressing these challenges requires a combination of empirical observation, theoretical understanding, and practical adjustments. By identifying potential issues and implementing preventive measures, brewers can improve their packaging processes.

Conclusion

Bottling and kegging are critical steps in the brewing process that require meticulous attention to detail and adherence to best practices. By applying principles of rationalism, empiricism, and the scientific method, brewers can optimize their packaging processes, ensuring that each batch of beer is well-preserved, properly carbonated, and ready for enjoyment.

The connections between these principles and the practical steps of bottling and kegging illustrate the blend of art and science in brewing, highlighting the continuous pursuit of excellence in the craft. Mastery of these techniques allows brewers to consistently produce high-quality beer that meets their standards and delights their customers.

> *To bottle or to keg is to seal a promise—the care taken not for oneself, but for the one who will lift the glass. The end, done rightly, redeems the whole*

**

Review Questions

True/False Questions

1. True or False: Sanitization is one of the most critical steps in bottling to prevent contamination.

2. True or False: Force carbonation involves adding priming sugar to the beer before sealing the keg.

3. True or False: The introduction of metal kegs revolutionized beer storage and transport in the 20th century.

4. True or False: Empirical studies have shown that different sanitizers can impact beer stability and flavor.

5. True or False: Proper bottle conditioning involves storing the bottled beer at room temperature for 1-2 weeks to allow for secondary fermentation.

Multiple Choice Questions

6. Which method of carbonation involves directly adding CO_2 to the kegged beer?
A) Natural carbonation
B) Priming carbonation
C) Force carbonation
D) Spontaneous carbonation

7. What is the ideal temperature range for bottle conditioning beer?
A) 32-34°F (0-1°C)
B) 46-50°F (8-10°C)
C) 68-72°F (20-22°C)
D) 75-80°F (24-27°C)

Brewer Vignettes

8. Brewer Vignette 1: You notice that your beer has developed off-flavors resembling cardboard after bottling. What is the likely cause, and how can you prevent it in the future?
A) Contamination during bottling; ensure thorough sanitization.
B) Over-carbonation; reduce the amount of priming sugar.
C) Oxidation; minimize oxygen exposure by purging bottles with CO_2 before sealing.
D) Improper storage; store bottles at cooler temperatures.

9. Brewer Vignette 2: As a brewer, you want to ensure consistent carbonation in your kegged beer. What practice should you follow?
A) Add priming sugar to the keg before sealing.
B) Monitor and adjust the serving pressure regularly to maintain desired carbonation levels.
C) Store the keg at room temperature for natural carbonation.
D) Use only whole hops during brewing to ensure proper carbonation.

10. Brewer Vignette 3: You are experiencing under-carbonation in your bottled beer. What could be the reason, and how can you fix it?
A) Not enough priming sugar; use a priming sugar calculator to determine the correct amount.
B) Excessive sanitization; reduce the amount of sanitizer used.
C) Improper storage; increase the storage temperature.
D) Using force carbonation; switch to natural carbonation methods.

**

Correct Answers
True/False Questions
1. True
2. False (Force carbonation involves directly adding CO_2 to the kegged beer.)

3. True
4, True
5. True

Multiple Choice Questions

6. C) Force carbonation
7. C) 68-72°F (20-22°C)

Brewer Vignettes

8. C) Oxidation; minimize oxygen exposure by purging bottles with CO2 before sealing.
9. B) Monitor and adjust the serving pressure regularly to maintain desired carbonation levels.
10. A) Not enough priming sugar; use a priming sugar calculator to determine the correct amount.

**

Beyond The Chapter

- Bamforth, C. W. (2003). Beer: Tap into the Art and Science of Brewing. Oxford University Press.
  - *Explores the processes of bottling and kegging and their impact on beer quality.*
- Palmer, J. (2006). How to Brew: Everything You Need To Know To Brew Beer Right The First Time. Brewers Publications.
  - *Provides practical advice on bottling and kegging techniques for homebrewers.*
- Noonan, G. (1996). Brewing Lager Beer: The Most Comprehensive Book for Home - And Microbrewers. Brewers Publications.
  - *Discusses the challenges and solutions for bottling and kegging lagers.*
- Kunze, W. (2014). Technology Brewing & Malting. VLB Berlin.
  - *Covers the technical aspects of bottling and kegging in commercial brewing.*

Weblinks

- Bottling Beer at Home - Brew Your Own: https://byo.com/?s=Bottling
- Kegging Your Homebrew - Northern Brewer: https://www.northernbrewer.com/pages/search-results-page?q=kegging
- The Science of Beer Bottling - Homebrew Academy: https://homebrewacademy.com/?s=Bottling
- Kegging and Serving Homebrew - Craft Beer & Brewing Magazine: https://beerandbrewing.com/search/?search=Kegging
- Guide to Bottling Homebrew - American Homebrewers Association: https://www.homebrewersassociation.org/?s=bottling
- Bottling and Kegging Best Practices - Homebrew Talk: https://www.homebrewtalk.com/search/8151489/?q=best+practices&o=relevance

These references offer comprehensive insights into bottling and kegging, serving as valuable resources for both homebrewers and professional brewers alike.

***

Chapter 10: The Silent Guardian—Cleanliness in Brewing

Cleanliness is the silent guardian of quality, ensuring each brew shines with precision and care.

Imagine entering a brewery where every detail, from the gleaming stainless steel tanks to the meticulously maintained hoses and fermenters, reflects a commitment to cleanliness and sanitation. This dedication is the backbone of brewing, ensuring that each batch of beer reaches the highest standards of quality and safety. The journey through the meticulous practices of cleaning and sanitation in brewing is as critical as the brewing process itself, embodying the blend of tradition, science, and innovation that defines this timeless craft.

Historical Origins

The history of cleaning and sanitation in brewing is rich and extensive, reflecting humanity's growing understanding of microbiology and its importance in food and beverage production. Early brewers in ancient Mesopotamia and Egypt had rudimentary methods for cleaning their equipment, relying on natural elements like boiling water and basic rinsing techniques. These practices were guided more by empirical observations than by a deep understanding of microbiological threats.

The story of beer began thousands of years ago, in the fertile valleys of Mesopotamia and along the banks of the Nile, where the first brewers, guided by intuition and observation, transformed humble grains into a beverage that became central to their communities. These early pioneers laid the foundation for a craft that would evolve through the ages, driven by the principles of rationalism, empiricism, and the scientific method.

Rationalism—Deductive Reasoning

Microbiology Informs Practical Decisions

Rationalism, with its emphasis on logical deduction and theoretical understanding, played a crucial role in the advancement of cleaning and sanitizing techniques. As knowledge about microorganisms expanded, brewers began to understand the necessity of creating surfaces that resisted microbial colonization. This theoretical approach led to the widespread adoption of materials like stainless steel, which is resilient, easy to clean, and resistant to corrosion by most common cleaners and sanitizers. The logical deduction that smooth, non-porous surfaces are less likely to harbor bacteria than rough, porous ones has driven the material choices in modern breweries. This rational approach has been integral to the development of modern brewing equipment and facilities.

Glass and Plastic Vessels

Glass, while still used in some homebrewing setups, has fallen out of favor in larger operations due to its brittleness and tendency to crack. Glass carboys, which are large, narrow-necked containers used for fermenting and storing liquids, are popular among homebrewers. Made of glass, carboys are non-reactive, preserving brew purity. They come in various sizes, typically from 1 to 6.5 gallons. However, in larger breweries, glass is less practical due to its fragility.

Plastic fermenters, though convenient and inexpensive, are generally avoided for long-term use as they can become scratched, providing a haven for microbes. Plastic carboys are lighter and less fragile than glass ones but can develop scratches over time, which can harbor bacteria and other microorganisms. Proper sanitation and handling are crucial to prevent contamination. Despite these drawbacks, plastic can still be useful for transferring wort and beer, provided it is properly cleaned and periodically replaced.

Empiricism—Inductive Reasoning

Observations Leading to Effective Choices

Empiricism, grounded in observation and experimentation, has allowed brewers to test their ideas and learn from their experiences. Over time, brewers empirically determined that certain cleaning agents were more effective than others.

- Mild cleaners like liquid hand-dishwashing detergents, which are soaps composed of natural fats and oils combined with alkaline substances, proved to be safe and effective for general use. These soaps contain surfactants, which reduce water's surface tension, enabling the emulsification and removal of organic residues like proteins and fats. However, soaps must be thoroughly rinsed off to prevent any residue from affecting beer flavor and head retention. While effective, soaps are often less efficient than specialized brewing cleaners and sanitizers designed for brewing environments.
- For heavy-duty cleaning tasks, more robust options like trisodium phosphate (TSP) became popular. TSP is a strong cleaning agent effective in removing heavy soils and residues from brewing equipment. It is particularly useful for tackling stubborn grime that regular detergents cannot remove. However, TSP must be handled carefully, following safety guidelines, and equipment should be thoroughly rinsed after use to remove any residue.
- Automatic dishwasher detergents, though powerful, were found to etch glass, leading to their selective use on metal surfaces. These detergents work exceptionally well on metal due to their strong cleaning action but can damage glass equipment over time.

Challenges with Complex Equipment

Equipment with hidden surfaces, such as counterflow wort coolers and certain filters, poses a significant challenge. A counterflow wort cooler is a device used to rapidly cool wort by having hot wort and cold water flow in opposite directions through adjacent pipes, enabling efficient heat exchange. These coolers have hidden surfaces that can harbor residues and microbes if not cleaned properly. They require scrupulous cleaning using powerful chemicals to ensure no residues or microbes remain.

To clean such complex equipment, brewers often employ methods like Clean-In-Place (CIP) systems and backflushing.

- Clean-In-Place (CIP) is a method used to clean the interior surfaces of pipes, vessels, process equipment, filters, and associated fittings without disassembly. This system circulates cleaning solutions through a closed loop to ensure thorough cleaning of complex equipment.
- Backflushing is a cleaning technique where cleaning solution is forced through equipment in the opposite direction of normal use to dislodge and remove deposits more effectively.

The complexity of cleaning such equipment underscores the importance of a thorough and systematic approach to brewery hygiene.

The Scientific Method—Advancing Cleaning Practices

Framework for Systematic Improvement

The scientific method has been indispensable in refining these practices. Brewers began with observations about what worked and what didn't, forming hypotheses about the best cleaning and sanitizing methods. Through controlled experiments, they tested these hypotheses, analyzed the results, and refined their techniques. This iterative process has led to significant advancements in brewery hygiene. The scientific method, with its structured approach to inquiry, has provided a framework for systematic improvement and innovation in cleaning and sanitation practices.

Cleaning Agents

The choice of cleaners is vital in maintaining the integrity and cleanliness of brewing equipment.

- Gentle cleaners such as liquid dish soaps are both safe and efficient for routine cleaning.
- For tougher cleaning jobs, trisodium phosphate (TSP) is highly effective, particularly for tackling stubborn grime.
- Automatic dishwasher detergents work exceptionally well on metal surfaces but should be used cautiously on glass due to their abrasive properties, which can cause etching.
- The most powerful cleaners are caustic soda (sodium hydroxide) and caustic potash (potassium hydroxide). These are strong alkaline cleaners used in the brewing industry to dissolve organic materials such as proteins and fats. They are unparalleled in their ability to dissolve protein residues but require strict safety protocols due to their corrosive nature. Protective gear such as gloves and goggles must be worn when handling these substances, and equipment must be thoroughly rinsed after use.
- Traditional cleaners like hypochlorite (bleach) have been used as sanitizers due to their effectiveness at killing microbes. However, bleach is prone to causing off-flavors and corrosion, making it less favored in modern commercial brewing. Its use requires careful handling and thorough rinsing to prevent residual chlorine from affecting the beer.

Sanitization Agents

Sanitization is the next critical step, ensuring that any cleaned surface is free from harmful microorganisms.

- Traditional sanitizers like chlorine and chlorinated TSP are effective at killing microbes but have fallen out of favor in commercial breweries due to their corrosive properties and potential to cause off-flavors in beer.

- Iodophors, sanitizers based on iodine, are a safer alternative for stainless steel and glass. They are effective at lower concentrations compared to chlorine-based sanitizers, killing microbes efficiently. Iodophors offer a visual indicator of their effectiveness by changing color when their antimicrobial activity is exhausted, making it easier for brewers to know when the solution needs to be refreshed.
- Newer oxygen-based sanitizers like chlorine dioxide and peroxyacetic acid are gaining popularity.
    - Chlorine dioxide is a gas used in public water treatment facilities; similar to bleach, it kills bacteria, viruses, and fungi but is less likely to form harmful byproducts.
    - Peroxyacetic acid is a modern, oxygen-based sanitizer known for its non-corrosive properties and suitability for no-rinse applications. It is gaining favor in commercial breweries for its effectiveness and safety, as it breaks down into harmless byproducts (water, oxygen, and acetic acid) after use.

Cleaning and Sanitizing Tools

The tools and techniques for cleaning and sanitizing brewing equipment are as varied as the equipment itself.

- Sponges are useful for gentle cleaning.
- Green pads—scrubbing tools made of abrasive material—are ideal for removing deposits from metal surfaces in the brewing process. They are effective but must be used carefully to avoid scratching equipment, especially on delicate surfaces.
- Nylon brushes are essential for cleaning bottles, small metal parts, and beer lines, reaching into crevices and ensuring thorough removal of residues.
- Pumps are indispensable for cleaning hoses, counterflow wort coolers, and filters, providing the pressure needed to flush out residues effectively.

Hot Side vs. Cold Side Cleaning

Hot Side Cleaning

The process of cleaning and sanitizing involves several steps, tailored to the specific needs of the hot and cold sides of brewing.

- On the hot side, where equipment comes into contact with hot wort, a green scrub pad with dishwashing liquid is often sufficient for cleaning.
- For stubborn residues, soaking equipment in a strong cleaning solution overnight can help loosen and remove deposits.

Cold Side Cleaning

Cold-side cleaning demands more caution to avoid scratching surfaces, which can harbor microbes.

- Carboys, which are large, narrow-necked containers used in brewing for fermenting and storing liquids, and hoses require thorough cleaning and inspection to ensure no deposits remain.
- Carboys can be made of glass or plastic.
  - Glass carboys are non-reactive and preserve brew purity but are heavy and fragile.
  - Plastic carboys are lighter and less fragile but can scratch easily, providing places for bacteria to hide.
- Proper sanitation and handling are crucial to prevent contamination on the cold side, where beer is more susceptible to microbial spoilage.

Clean-In-Place (CIP) Systems

A robust Clean-In-Place (CIP) system is invaluable for maintaining hygiene in hoses, racking tubes, and other intricate equipment. By circulating cleaning solutions through a closed loop, CIP systems effectively remove residues without disassembly.

This process involves several stages:

1. Backflushing to remove deposits: Cleaning solution is forced through equipment in the opposite direction of normal use to dislodge and remove deposits more effectively.
2. Circulating a cleaning solution to ensure all internal surfaces are contacted and cleaned thoroughly.
3. Thorough rinsing before sanitizing to remove any residual cleaning agents.

The evolution of CIP systems is a testament to the principles of rationalism, empiricism, and the scientific method. Early CIP systems were developed based on theoretical understanding of fluid dynamics and cleaning chemistry—a rational approach. Over time, empirical observations about the effectiveness of various cleaning cycles and chemicals refined these systems. The scientific method provided a structured approach to testing and validating these innovations, leading to the highly effective CIP systems used in modern breweries.

Cleaning Bottles and Kegs

Bottles and kegs also require meticulous cleaning and sanitizing.

Bottles

- Soaking bottles in a TSP solution helps remove labels and residues.
- Brushing and inspection ensure that all debris is removed from the interior and exterior surfaces.
- Sanitizing the bottles is crucial to eliminate any remaining microorganisms.
- For an effective sanitization method, bottles can be run through a dishwasher cycle with plain water, followed by a heat-dry cycle.

Kegs

- Kegs can be cleaned by hand, using appropriate cleaning agents and tools.
- CIP systems offer a more efficient solution for cleaning kegs, especially in larger operations.
- Regular disassembly of keg fittings ensures that all parts, including valves and seals, are thoroughly cleaned and sanitized.

Principles of Sanitizing

- Always follow manufacturer's directions for any cleaning and sanitizing agents.
- Re-sanitize any vessel or equipment that has been empty for more than a few days before use.
- Use boiling water where appropriate for an additional level of sanitation.

Conclusion

The history of cleaning and sanitation in brewing reflects humanity's growing understanding of microbiology and its importance in food and beverage production. Early brewers relied on rudimentary cleaning methods, guided more by empirical observations than by a comprehensive understanding of microbiological threats. The transition from these early practices to modern, scientifically informed methods showcases the evolution of brewing hygiene.

Driven by rationalism, empiricism, and the scientific method, the development of cleaning and sanitizing techniques has transformed brewing into a precise and reliable craft. By embracing these principles, brewers can continue to innovate and excel, creating beers that honor tradition while pushing the boundaries of what's possible.

Always remember that the true heart of brewing lies not just in recipes and techniques, but in the unwavering dedication to cleaning and sanitation that makes it all possible.

> *Cleanliness in brewing is the ethic before the craft—the quiet obligation to protect what will never speak, yet always be tasted.*

**

Review Questions

True/False Questions

1. True or False: Glass is the preferred material for large brewing operations due to its durability and ease of cleaning.

2. True or False: Rationalism, with its emphasis on logical deduction, has been crucial in the choice of materials for brewing equipment.

3. True or False: Chlorine-based sanitizers are highly recommended for no-rinse applications in brewing due to their effectiveness.

4. True or False: Empirical observations have shown that trisodium phosphate (TSP) is effective for heavy-duty cleaning in breweries.

5. True or False: Clean-In-Place (CIP) systems are used to clean brewing equipment without the need for disassembly.

Multiple Choice Questions

6. Which of the following is a disadvantage of using glass in large brewing operations?
A) It is prone to cracking
B) It resists microbial colonization
C) It is highly durable
D) It is easy to clean

7. Which modern sanitizer is gaining popularity for its non-corrosive properties and suitability for no-rinse applications?
A) Chlorine bleach
B) Chlorinated TSP
C) Peroxyacetic acid
D) Iodophors

Brewer Vignettes

8. Brewer Vignette 1: You are a brewer noticing that your fermenters are showing signs of microbial contamination. What steps should you take to ensure thorough sanitation?
A) Use only mild detergents for cleaning.
B) Apply a strong caustic soda solution, followed by proper rinsing.
C) Avoid using any cleaning agents to prevent damage.
D) Increase fermentation temperature to kill microbes.

9. Brewer Vignette 2: Your brewery is transitioning to using stainless steel equipment. How should you approach the cleaning and maintenance of this new equipment?
A) Use abrasive cleaners frequently to ensure cleanliness.
B) Implement a CIP system and use non-corrosive sanitizers like peroxyacetic acid.
C) Continue using glass and plastic for smaller operations.
D) Ignore scratches on the surfaces as they do not impact sanitation.

10. Brewer Vignette 3: After several batches, you find that your beer has developed off-flavors. You suspect the issue lies with your counterflow wort cooler. How should you address this problem?
A) Increase the hops to mask off-flavors.
B) Thoroughly clean and backflush the cooler using a CIP system.
C) Replace the wort cooler with a plastic fermenter.
D) Reduce the cleaning frequency to avoid equipment wear.

Correct Answers

True/False Questions

1. False (Glass is not the preferred material due to its brittleness and tendency to crack.)
2. True
3. False (Chlorine-based sanitizers are not recommended for no-rinse applications due to potential off-flavors.)
4. True
5. True

Multiple Choice Questions

6. A) It is prone to cracking
7. C) Peroxyacetic acid

Brewer Vignettes

8. B) Apply a strong caustic soda solution, followed by proper rinsing.
9. B) Implement a CIP system and use non-corrosive sanitizers like peroxyacetic acid.
10. B) Thoroughly clean and backflush the cooler using a CIP system.

**

Beyond The Chapter

- Palmer, J. (2017). How to Brew: Everything You Need To Know To Brew Beer Right The First Time. Brewers Publications.
  - *Explains the importance of cleaning and sanitation in brewing and provides practical tips.*
- Zainasheff, J., & Palmer, J. (2007). Brewing Classic Styles: 80 Winning Recipes Anyone Can Brew. Brewers Publications.
  - *Discusses best practices for cleaning and sanitation to ensure beer quality.*
- Papazian, C. (2003). The Complete Joy of Homebrewing. Avon Books.
  - *Provides an overview of essential cleaning and sanitation techniques for homebrewers.*

- White, C., & Zainasheff, J. (2010). Yeast: The Practical Guide to Beer Fermentation. Brewers Publications.
    - *Covers the impact of proper sanitation on yeast health and fermentation outcomes.*
- Grime, R. W. (1997). Sanitation in the Brewery. MBAA TQ (Master Brewers Association of the Americas Technical Quarterly).
    - *Offers detailed information on sanitation practices in commercial breweries.*
- Goldammer, T. (2008). The Brewer's Handbook: The Complete Book to Brewing Beer. Apex Publishers.
    - *A comprehensive guide on cleaning and sanitation procedures for brewers.*

WebLinks

- **BeerSmith Home Brewing Blog**: This resource covers various cleaning and sanitation practices essential for home brewers, offering practical tips and common pitfalls to avoid. https://beersmith.com/blog/
- **BREWHA Equipment Co**: Detailed guide on cleaning and sanitizing stainless brewing equipment, including the use of boiling for sanitization and maintenance tips. https://brewhaequipment.com/search?q=Sanitation&options%5Bprefix%5D=last
- **Homebrew Academy**: An in-depth look at various cleaning agents and sanitizers, including how to use them effectively in different brewing contexts. https://homebrewacademy.com/?s=Sanitation
- **Brewers Association**: Comprehensive guidelines on maintaining sanitation in professional brewing, from cleaning protocols to the latest in sanitizing technology. https://www.brewersassociation.org/?s=Sanitation
- **Craft Brewing Business**: Articles discussing advanced cleaning techniques and the science behind effective sanitation

in commercial brewing environments. https://www.craftbrewingbusiness.com/?s=Sanitation

- **MoreBeer**: Tips on cleaning and sanitizing homebrew equipment, including the benefits of different cleaning agents and sanitizers. https://www.morebeer.com/search?search=Sanitation
- **Brew Your Own**: Practical advice on keeping brewing equipment clean and sanitized, with a focus on homebrewing practices. https://byo.com/?s=Sanitation
- **Northern Brewer**: Guide to the best practices for cleaning and sanitizing brewing equipment, aimed at helping brewers avoid contamination. https://www.northernbrewer.com/pages/search-results-page?q=Sanitize

These resources provide a broad spectrum of information on cleaning and sanitizing brewing equipment, ranging from homebrewing tips to commercial brewing practices.

***

Chapter 11: Mastering Brewing—Solving Problems Through Rational Thought, Empirical Observation, and the Scientific Method

In brewing, every problem solved is a step towards mastery; through rational thought, empirical observation, and the scientific method, perfection is within reach.

Imagine encountering a brewing problem just as you think everything is going smoothly—a stuck fermentation, an off-flavor, or a clarity issue. Brewing beer is both an art and a science, filled with challenges that can affect the final product. Understanding these issues and knowing how to address them can turn a mediocre batch into an exceptional one. By applying principles of rationalism, empiricism, and the scientific method, brewers can systematically identify and solve these problems, ensuring consistent and high-quality results. This blend of theoretical knowledge and practical skills is crucial for mastering the craft of brewing. With the right approach, common pitfalls can be turned into learning opportunities, enhancing both the quality of the brew and the skill of the brewer.

Historical Context

Brewers have historically faced various challenges in producing high-quality beer. From ancient times to the modern era, methods and techniques for addressing common brewing problems have evolved significantly. Early brewers relied on trial and error and empirical observations to refine their processes. Over time, scientific advancements and a better understanding of fermentation, sanitation, and chemical interactions have led to more effective solutions.

Milestones such as the discovery of yeast's role in fermentation and the development of pasteurization have dramatically improved the ability to troubleshoot and resolve brewing issues. These historical advancements provide a foundation for modern brewing practices. By learning from past practices and integrating modern scientific principles, brewers today can achieve a level of consistency and quality that was unimaginable in earlier times.

Rationalism—Deductive Reasoning

The Role of Rationalism

Rationalism plays a crucial role in identifying and solving brewing problems. By applying deductive reasoning, brewers can develop theories and principles to anticipate and mitigate issues based on established knowledge. For example, understanding the theoretical underpinnings of yeast metabolism and fermentation dynamics allows brewers to predict potential problems and implement preventive measures. This top-down approach helps ensure that brewing practices are grounded in sound scientific principles.

Rationalism provides the framework for understanding complex brewing processes and applying this knowledge to practical situations. With a solid theoretical base, brewers can make informed decisions that enhance the efficiency and effectiveness of their brewing practices.

Empiricism—Inductive Reasoning

The Importance of Empiricism

Empiricism, grounded in careful observation and experimentation, has been equally important in refining troubleshooting techniques in brewing. Brewers have used inductive reasoning to gather data from countless experiments, identifying patterns and optimizing procedures. This bottom-up approach has led to practical innovations and best practices, ensuring that brewing processes are both scientifically sound and practically effective.

By continuously observing and experimenting, brewers can adapt their methods to improve outcomes. Empirical evidence provides the practical insights needed to complement theoretical knowledge.

This hands-on approach has been instrumental in advancing brewing techniques and achieving consistent results.

The Scientific Method—Structured Approach

Advancing Brewing Techniques

The scientific method provides a structured approach to improving brewing techniques. By following the steps of observation, hypothesis formation, experimentation to verify or falsify the hypothesis, data analysis with assessment, replicability, and peer review, brewers have made significant advances in understanding and optimizing their processes.

Specific case studies, such as the development of precise carbonation methods and the effects of different sanitizers on beer quality, illustrate how the scientific method has driven innovation in brewing. This systematic approach ensures that improvements are based on solid evidence and can be replicated by others in the brewing community. Applying the scientific method to brewing not only enhances the quality of the beer but also contributes to the broader body of knowledge in the field.

Fermentation Issues

Stuck Fermentation

One of the most frustrating issues a brewer can encounter is a stuck fermentation, where the yeast fails to fully ferment the sugars in the wort. This can result in a sweet, low-alcohol beer that lacks the intended balance and character. Addressing stuck fermentation requires a thorough understanding of yeast health, oxygenation, and temperature control.

By identifying the root causes and implementing targeted solutions, brewers can ensure a complete and robust fermentation process. Stuck fermentation can be a complex problem, but with the right knowledge and techniques, it can be effectively managed and resolved.

Causes and Solutions

- Low Yeast Health: Healthy yeast is essential for a vigorous and complete fermentation. If the yeast is old, under-pitched, or improperly stored, it may not have the vitality to finish the job. To address this, ensure you are using fresh yeast and pitching the correct amount. For high-gravity beers, consider using a yeast starter to increase cell count. Regularly monitoring yeast viability and vitality can prevent many fermentation issues.
- Insufficient Oxygen: Yeast requires oxygen to reproduce during the initial lag phase of fermentation. If the wort is not adequately oxygenated, yeast reproduction can be stunted, leading to a sluggish or stuck fermentation. Use an aeration stone—a device used to introduce oxygen into wort before fermentation, ensuring proper yeast growth and activity—or shake the wort vigorously before pitching the yeast to ensure sufficient oxygen levels. Proper oxygenation is crucial for healthy yeast growth and effective fermentation.
- Improper Fermentation Temperature: Yeast has an optimal temperature range for activity. If the fermentation temperature is too low, yeast activity can slow down or stop entirely. Conversely, if it is too high, yeast can produce off-flavors. Maintain fermentation temperatures within the yeast strain's recommended range, using temperature control systems like fermentation chambers or temperature wraps. Consistent temperature control is essential for maintaining yeast performance and preventing off-flavors.

Solutions for a Stuck Fermentation

If you find yourself with a stuck fermentation:

- Rouse the Yeast: Gently stir the fermenter to re-suspend the yeast in the wort.
- Adjust Temperature: Raise the temperature slightly within the yeast's optimal range to encourage activity.

- Add More Yeast: In severe cases, pitch a fresh, active yeast starter to re-inoculate the wort.

By understanding the underlying causes and implementing these solutions, brewers can effectively manage and resolve stuck fermentation issues.

Off-Flavors: Identification and Prevention

Off-flavors can ruin an otherwise great beer. Identifying and addressing these flavors is crucial for producing high-quality brews. Understanding the sources of off-flavors and implementing preventive measures can significantly improve the quality of your beer.

Common Off-Flavors and Solutions

- Diacetyl (Buttery Flavor): Produced by yeast during fermentation, diacetyl gives a buttery or butterscotch taste. To minimize diacetyl:
  - Ensure proper yeast health and adequate fermentation time.
  - Perform a diacetyl rest, where the fermentation temperature is raised slightly at the end of primary fermentation to help yeast reabsorb this compound.
- Acetaldehyde (Green Apple Flavor): An intermediate compound in ethanol formation, acetaldehyde imparts a green apple or raw pumpkin flavor. To avoid acetaldehyde:
  - Ensure proper pitching rates, oxygenation, and fermentation temperatures.
  - Allow the beer to fully ferment and condition before packaging.
- Phenolic (Medicinal Flavor): Phenolic off-flavors can taste medicinal, smoky, or spicy, often caused by wild yeast or bacteria contamination or high fermentation temperatures. To prevent phenolic flavors:
  - Maintain strict sanitation practices to prevent contamination.

- Control fermentation temperatures suitable for the yeast strain used.

Clarity and Haze

Achieving clarity is a key aspect of many beer styles, and unwanted haze can detract from the beer's appearance and quality. Several factors can contribute to haze, including proteins, polyphenols, and yeast in suspension. Addressing these issues can improve the visual appeal and quality of your beer.

Protein and Polyphenol Haze

- Chill Haze: Becomes apparent when the beer is chilled, caused by proteins and polyphenols binding together. Polyphenols are compounds in malt and hops that influence beer's flavor, stability, and clarity. They contribute to astringency and bitterness and have antioxidant properties but can form complexes with proteins, causing haze.

Solutions for Haze

- Fining Agents: Using fining agents during the brewing process can help precipitate haze-forming particles:
    - Irish Moss: A type of red seaweed added during the last 15 minutes of the boil. It helps coagulate proteins and suspended solids, improving clarity.
    - Whirlfloc Tablets: Made from Irish moss and purified seaweed, these tablets are added during the last 15 minutes of the boil to enhance protein coagulation and settling.
    - Gelatin: Derived from animal collagen, gelatin is dissolved in warm water and added to cold beer. It binds to proteins and haze-forming compounds, causing them to settle.
- Cold Crashing: Lowering the temperature of the beer near freezing before packaging causes proteins and yeast to settle out, improving clarity.

Yeast in Suspension

- Filtration: Using a filtration system removes yeast and other particulates.
- Cold Storage: Allowing the beer to condition at low temperatures enables yeast to settle naturally.

Carbonation Problems

Carbonation levels are crucial for the mouthfeel and overall perception of beer. Both under-carbonation and over-carbonation can affect the enjoyment of your brew. Proper carbonation techniques are essential for achieving the desired sensory experience.

Solutions for Carbonation Issues

- Under-Carbonation:
    - Check Priming Sugar: Ensure accurate calculations and measurements.
    - Seal Bottles Properly: Inadequate sealing can allow CO_2 to escape.
    - Consider Re-Priming: If necessary, re-prime bottles or use force carbonation for kegged beer.
- Over-Carbonation:
    - Measure Priming Sugar Accurately: Avoid adding too much sugar.
    - Mix Thoroughly but Gently: Ensure even distribution without introducing oxygen.
    - Sanitation: Prevent contamination by wild yeast or bacteria that can cause additional fermentation.

Practical Tips

Maintaining consistent brewing practices is crucial to avoiding common problems and ensuring high-quality beer. Consistency helps brewers identify and isolate variables that can affect the final product.

Mashing

- Grain Crush: Ensure consistency to promote even conversion of starches to sugars.
- Temperature Control: Maintain a steady mash temperature for proper enzyme activity.

Boiling

- Boil Duration: Follow recommended times for hop utilization and sterilization.
- Avoid Boil-Overs: A boil-over occurs when foam rises and spills over the kettle, causing loss of wort and potential safety hazards. Monitor the boil and adjust heat as necessary.

Fermentation

- Yeast Health: Use fresh, viable yeast and pitch at appropriate rates.
- Temperature Management: Maintain consistent fermentation temperatures suited to the yeast strain.
- Sanitation: Use sanitized equipment to prevent contamination.

Packaging

- Priming Sugar Measurement: Accurately calculate and measure to achieve desired carbonation.
- Sanitize Equipment: Ensure bottles, caps, and kegging equipment are thoroughly sanitized.

Ongoing Monitoring

Regularly monitoring brewing conditions is essential to catch and address issues early.

- Temperature: Use reliable thermometers for mash, boil, and fermentation stages.

- pH Levels: Measure and adjust the pH of the mash and wort for optimal enzyme activity and fermentation health.
- Gravity Readings: Use a hydrometer or refractometer to track fermentation progress and determine alcohol content.

Challenges and Solutions

Providing real-world examples of common brewing problems illustrates how experienced brewers have used rational, empirical, and scientific methods to address them.

- Example 1: A homebrewer notices a recurring acetaldehyde flavor in their beer. Through empirical observation, they identify that premature bottling is the issue. By extending the fermentation period and ensuring complete fermentation, the problem is resolved.
- Example 2: A commercial brewery faces protein haze issues. Using deductive reasoning, they hypothesize that insufficient protein rest during mashing is the cause. Adjusting the mash schedule to include a longer protein rest results in clearer beer.
- Example 3: A craft brewer encounters over-carbonation in bottled beer. Through a scientific approach, they conduct experiments with different priming sugar levels and mixing techniques.

Attending workshops and conferences—events where brewers learn from industry experts and participate in practical demonstrations—provides valuable insights. Peer review and community feedback help refine their process, leading to consistently carbonated beer.

Conclusion

In this chapter, we explored common brewing problems and their solutions, emphasizing the importance of rationalism, empiricism, and the scientific method. By understanding the historical context, applying deductive and inductive reasoning, and utilizing the scientific method, brewers can effectively troubleshoot and improve their brewing processes.

Exploring these aspects of brewing problems and solutions, we gain a comprehensive understanding of how to refine our craft. Whether through the application of rational theories, empirical observations, or the scientific method, the journey of troubleshooting and perfecting beer is a testament to the enduring blend of art and science in brewing.

> *To solve a brewing problem is not merely to rescue a batch, but to honor the deeper calling of the brewer: to think clearly, observe honestly, and trust that knowledge—earned through trial and truth—can transform error into excellence.*

**

Review Questions

True/False Questions

1. True or False: Stuck fermentation occurs when the yeast fails to fully ferment the sugars in the wort.

2. True or False: Low yeast health is one of the causes of stuck fermentation.

3. True or False: Proper oxygenation of wort is not necessary for yeast reproduction.

4. True or False: Phenolic off-flavors in beer can taste medicinal and are often caused by wild yeast or bacteria contamination.

5. True or False: Using fining agents like Irish moss can help improve beer clarity by reducing protein and polyphenol haze.

Multiple Choice Questions

6. Which of the following is a common cause of stuck fermentation?
A) Over-pitching yeast
B) Proper oxygenation
C) Low yeast health
D) Adequate fermentation temperature

7. What is a recommended practice to address protein haze in beer?
A) Use an aeration stone
B) Perform a diacetyl rest
C) Cold crashing the beer
D) Increase priming sugar

Brewer Vignettes

8. Brewer Vignette 1: Imagine you are a brewer dealing with a stuck fermentation. Your yeast seems to be inactive. What steps would you take to address this issue?
A) Raise the temperature slightly and stir the fermenter.
B) Ignore the issue and continue brewing.
C) Add more hops to the fermenter.
D) Remove some of the wort to reduce volume.

9. Brewer Vignette 2: As a brewer, you notice that your beer has developed a buttery flavor. What might be the cause and how would you address it?
A) High fermentation temperature; ignore the flavor.
B) Diacetyl presence; perform a diacetyl rest.
C) Low oxygenation; increase priming sugar.
D) Chlorine in water; use Campden tablets.

10. Brewer Vignette 3: You find that your beer has a cloudy appearance due to yeast in suspension. How would you improve clarity?
A) Ignore the haze and package the beer.
B) Boil the beer to remove yeast.
C) Use a filtration system or cold storage to settle yeast.
D) Add more yeast to the fermenter.

**

Correct Answers
True/False Questions
1. True
2. True
3. False (Correct Answer: Proper oxygenation of wort is necessary for yeast reproduction.)
4. True
5. True
Multiple Choice Questions
6. C) Low yeast health
7. C) Cold crashing the beer
Brewer Vignettes
8. A) Raise the temperature slightly and stir the fermenter.
9. B) Diacetyl presence; perform a diacetyl rest.
10. C) Use a filtration system or cold storage to settle yeast.

**

Beyond The Chapter

- Bamforth, Charles W. Beer: Tap into the Art and Science of Brewing. Oxford University Press, 2009.
  - *Explores common brewing problems and offers solutions to improve beer quality.*
- Fix, George. Principles of Brewing Science: A Study of Serious Brewing Issues. Brewers Publications, 1999.

  - *Discusses scientific principles and troubleshooting techniques for brewing issues.*
- Papazian, Charlie. The Complete Joy of Homebrewing. HarperCollins, 2014.
  - *Provides practical advice on identifying and resolving common brewing problems.*
- Palmer, John J. How to Brew: Everything You Need to Know to Brew Great Beer Every Time. Brewers Publications, 2017.
  - *Offers troubleshooting tips and solutions for various brewing challenges.*

Weblinks

- Homebrew Troubleshooting Guide - Brew Your Own: https://byo.com/?s=trouble+shooting
- Common Off-Flavors in Beer and How to Avoid Them - American Homebrewers Association: https://www.homebrewersassociation.org/?s=off+flavors
- Troubleshooting Off-Flavors - Craft Beer & Brewing Magazine: https://beerandbrewing.com/search/?search=off+flavors
- Common Homebrew Issues and How to Fix Them - Homebrew Talk: https://www.homebrewtalk.com/threads/common-homebrew-issues-and-how-to-fix-them.563246/

These references offer comprehensive insights into the problems and solution in the brewing process, serving as valuable resources for both homebrewers and professional brewers.

***

Chapter 12: The Path to Perfection—Continuous Improvement in Brewing

Continuous improvement in brewing lies at the intersection of rationalism, empiricism, and the scientific method, transforming each batch into a step towards perfection.

Imagine the satisfaction of pouring a perfectly brewed beer that you have refined over countless batches. Continuous improvement is essential in brewing to achieve consistently high-quality results. This process involves a commitment to refining techniques, experimenting with new methods, and learning from each brewing experience. Key concepts such as meticulous record-keeping, strategic experimentation, and continuous learning are integral to this process. These concepts tie into the overarching themes of rationalism, empiricism, and the scientific method, providing a structured approach to enhancing brewing techniques. By embracing continuous improvement, brewers can elevate their craft, ensuring each batch of beer surpasses the previous one in quality and flavor.

Historical Context

The tradition of tracking and improving brewing processes dates back centuries. Early brewers relied on oral traditions and rudimentary records to pass down their brewing knowledge. These methods, though effective to a degree, often lacked the precision needed for consistent results. Over time, more systematic methods emerged, particularly in medieval monasteries where monks meticulously documented their brewing practices. These early records laid the foundation for modern brewing logs and the application of scientific principles during the Industrial Revolution, which brought about significant advancements in brewing technology and methodology.

Rationalism and empiricism have profoundly influenced brewing practices. Rationalism provided the theoretical foundation for understanding fermentation and ingredient interactions, while empiricism drove practical improvements through observation and experimentation. This historical evolution demonstrates the importance of combining theoretical knowledge with practical application to achieve continuous improvement in brewing.

Rationalism—Deductive Reasoning

Rationalism plays a crucial role in improving brewing techniques by applying top-down deductive reasoning. Theoretical frameworks and established principles guide systematic improvements in brewing practices. For example, understanding the enzymatic activity in mashing and the impact of different yeast strains on fermentation is based on rationalist principles. By starting with a theoretical hypothesis, such as the expected outcome of using a specific malt type, brewers can test and refine these hypotheses through practical application.

Applying Theoretical Knowledge

Brewing improvements often begin with a theoretical foundation:

- Enzymatic Activity: Knowing how enzymes like alpha-amylase and beta-amylase break down starches into fermentable sugars allows brewers to optimize mash temperatures and times to achieve the desired sugar profile.
- Yeast Strain Selection: Understanding how different yeast strains metabolize sugars and produce flavor compounds helps brewers select the best yeast for each beer style. For instance, certain yeast strains produce more esters, contributing fruity notes desirable in ales.
- Hop Utilization: Theoretical knowledge about the isomerization of alpha acids during boiling guides decisions on hop schedules—the timing and quantity of hop additions during brewing, influencing bitterness, aroma, and flavor.

By applying rational thought, brewers can anticipate and mitigate potential brewing problems before they occur. This top-down approach ensures that brewing practices are grounded in sound scientific principles and helps prevent common issues such as off-flavors or inconsistent fermentation.

Empiricism—Inductive Reasoning

Empiricism is vital in brewing improvement, emphasizing a bottom-up approach through inductive reasoning. Brewers gather empirical data through observations and experiments to refine their techniques. By systematically experimenting with different variables and recording the outcomes, brewers can identify the best practices for each step of the brewing process.

Practical Experimentation

Examples of empirical practices include:

- Water Chemistry Adjustments: Experimenting with the pH and mineral content of brewing water to observe how it affects mash efficiency and flavor extraction.
- Mash Temperature Variations: Adjusting mash temperatures to see how it affects the fermentable sugar profile and, consequently, the beer's body and alcohol content.
- Fermentation Conditions: Testing different fermentation temperatures to determine their impact on yeast performance and flavor development.

Through repeated trials and careful analysis of the results, brewers can determine the optimal conditions for their recipes. Empiricism ensures that brewing practices are continually refined based on real-world observations and outcomes, leading to continuous improvement.

The Scientific Method—Structured Approach

The scientific method provides a structured approach to improving brewing practices. The steps include:

1. Observation: Monitor brewing processes and outcomes to identify areas for improvement.
2. Hypothesis Formation: Develop theories about process improvements based on rational understanding.
3. Experimentation: Empirically test different variables and conditions to verify or falsify the rational hypothesis.
4. Data Analysis: Analyze the empirical results to draw conclusions and refine brewing practices.
5. Replicability: Ensure that the results can be replicated using the same methods and procedures.
6. Peer Review: Share findings with the brewing community for feedback and validation.

Applying the Scientific Method

Start with an observation—for example, noticing a recurring off-flavor in your beer. Form a rational hypothesis about its cause, such as contamination from inadequate sanitization. Conduct empirical experiments to verify or falsify your rational hypothesis, like improving your sanitization process. Analyze the empirical data to determine if the change resolved the issue. Peer review and community feedback can further validate your findings and refine your methods.

The integration of rationalism, empiricism, results in the scientific method. Using the scientific method brewers can systematically enhance their techniques and consistently produce high-quality beer. Specific case studies, such as optimizing fermentation temperatures or experimenting with hop schedules, illustrate how the scientific method leads to significant advancements in brewing.

Practical Applications and Techniques

Record Keeping

Meticulous record-keeping is fundamental to improving brewing processes. Detailed brewing logs help track ingredients, temperatures, times, and deviations from the plan. Analyzing these logs can reveal patterns and insights, allowing brewers to replicate successful batches and troubleshoot issues.

What to Record

- Ingredients: Document the types, quantities, and sources of all ingredients used, including malt, hops, yeast, and any adjuncts.
- Temperatures and Times: Record all stages from mashing to fermentation, noting specific temperatures and durations for each step.
- Deviations: Note any changes from the original plan, such as adjustments made during the brewing process.
- Process Notes: Include observations about mash efficiency, boil vigor, and fermentation activity.
- Tasting Notes: Use standardized tasting sheets to describe the aroma, flavor, mouthfeel, and appearance of your beer, noting how these elements change over time.

Analysis and Patterns

By analyzing your brewing logs, you can identify patterns and correlations between specific variables and the final beer quality. For example:

- Mash Temperature Impact: Noticing that a mash temperature of 152°F consistently produces a smoother mouthfeel compared to higher temperatures.
- Yeast Performance: Observing that a particular yeast strain ferments more vigorously and produces preferred flavor profiles under certain conditions.

This data-driven approach is invaluable for making informed adjustments and achieving desired outcomes. Over time, you can build a personalized database of brewing knowledge, enhancing your ability to produce high-quality beer consistently.

Modern Advancements

Recent advancements in technology and methodology have revolutionized brewing experimentation. Modern brewers use rationalism, empiricism, and the scientific method to innovate and refine techniques. These innovations have significantly improved the precision and efficiency of brewing processes.

Technological Advancements

- Brewing Software: Digital tools used by brewers to design recipes, calculate efficiency, predict outcomes, and record brewing data. Examples include BeerSmith and Brewer's Friend. These programs allow precise adjustments to ingredients and processes.
- Automated Systems: Equipment that provides precise control over brewing conditions, such as temperature, pressure, and fermentation time. Systems like the Grainfather or Brewie automate many brewing steps, increasing consistency.
- Data Analytics: The process of examining brewing data to identify patterns and correlations that can optimize brewing operations and improve product quality. By analyzing data from multiple batches, brewers can identify trends and make informed decisions.

Cutting-Edge Practices

- Advanced Sensors: Real-time monitoring of fermentation parameters like temperature, pH, and gravity using sensors connected to digital platforms.
- Artificial Intelligence (AI): Employing AI algorithms to predict fermentation outcomes, flavor profiles, and optimize recipes based on historical data.

By leveraging modern advancements, brewers can push the boundaries of their craft and continuously improve their techniques.

Flavor Development

Keeping detailed tasting notes for each batch of beer is equally important. Tasting notes are detailed descriptions of the aroma, flavor, mouthfeel, and appearance of beer, used to track changes over time and identify areas for improvement. Document how these elements change, especially during bottle conditioning or keg aging.

Understanding Beer Character

This practice helps you understand how different factors influence the beer's character:

- Ingredient Impact: How specific malts affect color and flavor.
- Process Influence: The effect of fermentation temperature on ester production.
- Aging Effects: How flavors develop and mellow over time.

For example, noting the progression of hop aroma during aging can inform future dry hopping schedules. Detailed tasting notes serve as a valuable reference for refining recipes and improving the sensory qualities of your beer.

Identifying Issues

Tasting notes can also help pinpoint specific issues:

- Off-Flavors: Identifying unwanted flavors like diacetyl (buttery) or acetaldehyde (green apple) can indicate fermentation problems.
- Balance: Assessing whether the beer is too bitter or sweet guides adjustments in hop additions or mash profiles.

By comparing notes across multiple batches, you can identify recurring problems and their potential solutions. This systematic approach to tasting and analysis is crucial for continuous improvement.

Experimentation and Innovation

Recipe Development

Brewing is a creative process that benefits from experimentation with different ingredients. Try using various malts, hops, yeast strains, and adjuncts to create unique flavor profiles.

- Malts: Experiment with specialty malts like Munich or Crystal to add depth and color.
- Hops: Try new hop varieties like Mosaic or Galaxy to introduce complex tropical fruit or citrus notes.
- Yeast Strains: Use different yeast strains to explore variations in ester and phenol production.

Use brewing software to design and refine your recipes, adjusting variables to achieve the desired outcomes. Historical examples, such as the development of the India Pale Ale (IPA) style through high hop levels highlight the benefits of ingredient experimentation.

Techniques and Styles

In addition to ingredient experimentation, try different brewing techniques:

- Dry Hopping: Adding hops during or after fermentation to enhance aroma without increasing bitterness.
- Barrel Aging: The process of maturing beer in wooden barrels, which imparts complex flavors from the wood and any previous contents of the barrel. This technique enhances the depth, character, and aroma of the beer.
- Sour Beers: Beers characterized by a tart and acidic flavor, often produced through mixed fermentation with wild yeast—naturally occurring yeast strains that contribute unique flavors and characteristics—and bacteria.

For example, adding cherries to a sour beer can create a complex, fruity character. Modern examples, such as the rise of New

England IPAs with their hazy appearance and juicy hop character, demonstrate the impact of innovative brewing techniques.

Collaboration

Collaborating with other brewers is a valuable way to share knowledge, learn new methods, and gain feedback on your brewing process.

Working Together

- Collaboration: Working jointly with other brewers on projects to share knowledge, learn new methods, and gain feedback. This fosters community and innovation in brewing.
- Local Clubs: Join local homebrew clubs to connect with fellow brewers.
- Online Communities: Participate in forums and social media groups dedicated to brewing.

These platforms provide opportunities to exchange ideas, troubleshoot issues, and stay informed about the latest trends and techniques. Collaboration fosters a sense of community and accelerates the collective improvement of brewing skills.

Joint Projects

Working on joint brewing projects can inspire creativity and innovation:

- Recipe Sharing: Co-develop recipes to blend different brewing styles and preferences.
- Group Experiments: Conduct parallel brewing experiments to compare results under different conditions.
- Mentorship: Experienced brewers can guide novices through complex techniques.

Historical collaborations, such as the cooperative efforts of monastic brewers in medieval Europe, demonstrate the long-standing benefits of shared brewing knowledge and resources.

Continuous Learning

Brewing Books and Journals

Stay updated with the latest research, trends, and techniques by reading widely from brewing books, magazines, and academic journals. These resources provide in-depth knowledge on various aspects of brewing, from ingredient science to advanced techniques.

- Recommended Books:
  - "How to Brew" by John Palmer
  - "The Brewmaster's Bible" by Stephen Snyder
  - *"Brewing Classic Styles"* by Jamil Zainasheff and John Palmer

Latest Trends

Keeping up with the latest trends in the brewing industry can introduce you to new styles, ingredients, and brewing methods:

- New Beer Styles: Explore emerging styles like Brut IPAs or Pastry Stouts.
- Innovative Ingredients: Experiment with novel adjuncts like exotic fruits, spices, or even coffee.
- Sustainability Practices: Adopt environmentally friendly brewing methods, such as water conservation or renewable energy use.

Embracing new trends and techniques can enhance your brewing portfolio and keep your offerings fresh and exciting.

Workshops and Conferences

Skill Development

Attend brewing workshops, conferences, and seminars to learn from industry experts. These events offer practical demonstrations, technical sessions, and networking opportunities with fellow brewers.

- American Homebrewers Association's Homebrew Con: An annual conference featuring educational seminars and workshops.
- Craft Brewers Conference: Offers insights into professional brewing practices and industry trends.

Participating in these events can significantly enhance your brewing expertise and provide new perspectives on the craft.

Networking

Networking—building relationships with other brewers and industry professionals—is crucial:

- Engage with Speakers: Ask questions and participate in discussions during sessions.
- Join Discussions: Participate in forums and workshops to share your experiences.
- Collaborative Opportunities: Find potential partners for future brewing projects.

Networking fosters a sense of community and support within the brewing industry. The connections you make can significantly enhance your brewing knowledge and open doors to new possibilities.

Thematic Connection

Continuous improvement in brewing is deeply connected to rationalism, empiricism, and the scientific method. These principles guide systematic enhancements and drive innovation in brewing practices. By applying these methods, brewers can achieve consistently high-quality results and push the boundaries of what is possible in brewing, a testament to the enduring blend of art and science in this beloved craft.

> *Perfection is not a place one reaches but a path one honors —with patience, reflection, and the humble willingness to begin again. In brewing, as in life, mastery belongs not to*

those who never err, but to those who improve with care, batch by batch.

**

Review Questions

True/False Questions

1. True or False: Continuous improvement in brewing involves refining techniques, experimenting with new methods, and learning from each brewing experience.

2. True or False: Rationalism does not play a significant role in brewing improvement.

3. True or False: Detailed brewing logs help track ingredients, temperatures, times, and deviations from the plan.

4. True or False: Empiricism involves gathering empirical data through observation and experimentation to refine brewing techniques.

5. True or False: Modern advancements in brewing technology have made record-keeping obsolete.

Multiple Choice Questions

6. What is the role of rationalism in brewing improvement?
A) It emphasizes the importance of trial and error.
B) It provides the theoretical foundation for understanding brewing processes.
C) It disregards the importance of theoretical knowledge.
D) It focuses solely on practical application.

7. Which practice is essential for continuous improvement in brewing?
A) Ignoring historical brewing methods.
B) Avoiding collaboration with other brewers.
C) Keeping meticulous brewing logs.
D) Limiting experimentation with new ingredients.

Brewer Vignettes

8. Brewer Vignette 1: You notice that your beer has developed a recurring off-flavor. Describe the steps you would take using the scientific method to identify and address the issue.
A) Ignore the issue and continue brewing.
B) Use deductive reasoning to develop a hypothesis, conduct experiments, and analyze the results.
C) Make random changes to the brewing process without recording the outcomes.
D) Rely solely on historical brewing methods without experimentation.

9. Brewer Vignette 2: As a brewer, you want to enhance the fruity character of your IPA. What steps could you take to achieve this?
A) Ignore tasting notes and continue with the same recipe.
B) Experiment with different hop varieties and document the results.
C) Avoid using new ingredients and stick to traditional methods.
D) Limit the analysis of brewing logs to past failures only.

10. Brewer Vignette 3: You are collaborating with another brewer to create a new beer style. What approach would you take to ensure a successful outcome?
A) Keep the brewing process a secret and avoid sharing knowledge.
B) Experiment with different brewing techniques and ingredients, documenting each step.
C) Follow only one brewer's method without any collaboration.
D) Avoid attending workshops and conferences for new ideas.

**

Correct Answers

True/False Questions

1. True
2. False (Correct Answer: Rationaism is essential for understanding theory and for deductive assessments.)
3. True
4. True
5. False (Correct Answer: The importance of meticulous record keeping has, and always will be the hallmark for improvement.)

Multiple Choice Questions

6. B) It provides the theoretical foundation for understanding brewing processes.
7. C) Keeping meticulous brewing logs.

Brewer Vignettes

8. B) Use deductive reasoning to develop a hypothesis, conduct experiments, and analyze the results.
9. B) Experiment with different hop varieties and document the results.
10. B) Experiment with different brewing techniques and ingredients, documenting each step.

**

Beyond The Chapter

- Bamforth, Charles W. Beer: Tap into the Art and Science of Brewing. Oxford University Press, 2009.
    - *Provides insights and tips for improving brewing techniques and beer quality.*
- Fix, George. Principles of Brewing Science: A Study of Serious Brewing Issues. Brewers Publications, 1999.
    - *Explores scientific principles and advanced techniques to enhance brewing.*
- Papazian, Charlie. The Complete Joy of Homebrewing. HarperCollins, 2014.

    - *Offers practical tips and advice for homebrewers to improve their brewing skills.*
- Palmer, John J. How to Brew: Everything You Need to Know to Brew Great Beer Every Time. Brewers Publications, 2017.
    - *Provides comprehensive guidance on improving various aspects of the brewing process.*
- Zainasheff, Jamil. Brewing Classic Styles. Brewers Publications, 2007.
    - *Discusses techniques for refining brewing practices to achieve classic beer styles.*
- White, Chris, and Jamil Zainasheff. Yeast: The Practical Guide to Beer Fermentation. Brewers Publications, 2010.
    - *Covers advanced yeast management techniques to improve fermentation outcomes.*
- Homebrew Talk: An online community with valuable discussions and resources.
    - *Offers peer advice and tips from experienced homebrewers to enhance brewing practices.*
- American Homebrewers Association: Offers articles, recipes, and educational materials.
    - *Provides resources and educational materials for improving brewing techniques.*

Weblinks

- Tips to Improve Your Homebrew - Homebrew Academy: https://homebrewacademy.com/?s=Tips
- Homebrewing Techniques and Tips - American Homebrewers Association: https://www.homebrewersassociation.org/?s=Tips
- Expert Homebrew Tips - Brew Your Own: https://byo.com/?s=Tips
- Common Homebrewing Mistakes and How to Avoid Them - Brew Cabin: https://www.brewcabin.com/homebrewing-mistakes/

- Pro Tips for Homebrewing - Craft Beer & Brewing Magazine: https://beerandbrewing.com/search/?search=Tips

These references provide a plethora of tips for improving your brewing and serving as valuable resources for both homebrewers and professional brewers alike.

***

Chapter 13: The Dance of Time, Wood, and Science—Mastering Barrel Aging

Barrel aging transforms beer through a dance of time, wood, and science, where patience and precision create layers of complexity.

Imagine the rich, complex aromas wafting from an oak barrel as beer slowly matures within. Barrel aging is a sophisticated technique that introduces new dimensions to beer, adding layers of complexity that are difficult to achieve through conventional brewing methods alone. This process requires patience, skill, and a deep understanding of how different elements interact over time. Historically, barrel aging has been used for centuries to enhance the flavors of beer, wine, and spirits, creating a rich tapestry of tastes and aromas. By drawing on the principles of rationalism, empiricism, and the scientific method, brewers can master the art of barrel aging and wood treatment, transforming their brews into exceptional products.

Barrel aging is a fascinating and historical method used to enhance the flavor and complexity of beer. This process not only imparts unique characteristics to the brew but also connects modern brewers to ancient traditions. The choice of barrel, the wood it is made from, and its previous contents all play crucial roles in the final flavor profile of the beer. This chapter delves into the intricate process of barrel aging and wood treatment, exploring how these elements contribute to the art and science of brewing.

Key components essential to understanding this process include tannins, which add astringency and complexity; vanillin, contributing sweet, creamy notes; and various oak varieties like American, French, and Hungarian, each imparting distinct flavors. These elements interact in complex ways to create a unique profile for each batch of barrel-aged beer.

Historical Context

Barrel aging has a rich history, tracing back to ancient civilizations where wooden vessels were used for storage and transportation. The Romans and Greeks utilized barrels for storing wine and other liquids, but it was the Celts who perfected the art of barrel making and aging. Over centuries, these techniques evolved, influenced by cultural exchanges and technological advancements.

Before the widespread use of wooden barrels, ancient brewers used amphorae—ancient clay pots with two handles and a narrow neck—for fermenting, aging, and storing beer. These vessels, made from porous clay material, imparted distinct flavors to the beer due to their interaction with the liquid. Modern craft brewers are reviving the use of amphorae for the unique characteristics they bring to the beer, connecting contemporary practices with ancient traditions.

Significant milestones include the transition from amphorae to barrels made from wooden staves bound by metal hoops. This shift also saw the introduction of oak as the preferred wood and the development of charring techniques to enhance flavor. Figures like Louis Pasteur, who studied the microbial stability of aged beverages, played crucial roles in refining these methods. Pasteur's research into fermentation and spoilage prevention laid the groundwork for modern barrel aging techniques, ensuring that beers aged in barrels were both flavorful and stable.

Rationalism and empiricism profoundly influenced the evolution of barrel aging. Theoretical approaches guided the selection of wood types and aging durations, while empirical observations helped optimize these practices for desired outcomes. This combination of theory and practice continues to drive advancements in barrel aging, allowing brewers to perfect their craft.

Types of Barrels

American Oak

American oak barrels, made from *Quercus alba*, commonly known as white oak, are renowned for their bold and distinctive flavors, making them a popular choice among brewers. Native to the eastern and central regions of North America, such as Missouri, Kentucky, and Tennessee, American oak has a more porous grain structure compared to French and Hungarian oak. This porosity allows for faster extraction of flavors, making the barrel aging process more efficient.

The heavy charring of these barrels caramelizes the wood sugars, adding a rich, smoky depth to the beer. This charring process also creates a porous structure in the wood, allowing the beer to interact more intimately with the barrel, extracting flavors such as vanilla, coconut, and sweet spices.

- Vanillin: A flavor compound found in oak that imparts vanilla-like aromas and flavors to beer. It adds a sweet, creamy note, particularly appealing in darker, richer beer styles like stouts and porters.
- Lactones: Compounds in oak that contribute to coconut and woody aromas in barrel-aged beer. American oak is high in lactones, making it a popular choice for a pronounced woody character.

Historically, American oak has been used extensively in bourbon aging, and reusing these barrels for beer infuses the brew with residual bourbon flavors, adding an extra layer of complexity. The barrels' previous use enhances the beer with notes of caramel, toffee, and vanilla, creating a rich and nuanced final product.

Scientific analysis reveals that chemical compounds such as vanillin and lactones contribute significantly to the flavor profile, providing sweet, creamy notes and a pronounced woody character. Understanding these compounds helps brewers predict and control the flavor development during the aging process, ensuring consistent and desirable results.

French Oak

French oak barrels, made from *Quercus robur*, known as European or English oak, offer a more subtle and refined flavor profile compared to American oak. These barrels are known for their fine grain and tighter pores, which result in a slower transfer of flavors into the beer. The tighter grain structure means flavors are extracted more gradually, allowing for greater control over the aging process.

French oak is known for imparting subtle and refined flavors such as gentle tannins, soft vanilla, and nuanced spice notes to beer. The elegance of French oak can enhance the complexity of the beer without overwhelming its inherent characteristics, providing a balanced and sophisticated finish. French oak is often used for aging lighter, more delicate beer styles.

- Ellagitannins: Compounds found in French oak that add gentle astringency and complexity to beer. They contribute to a refined, wine-like character.

French oak barrels are commonly used in winemaking, particularly for aging fine wines, and this practice influences the beer when these barrels are repurposed. Native to Europe, especially regions like France, Germany, and the UK, French oak's tight grain and slower flavor transfer make it a prized material in both the winemaking and brewing industries.

Scientific analysis shows that French oak contains compounds like ellagitannins, which interact with the beer to create sophisticated flavors, adding depth and refinement. Brewers can leverage this knowledge to tailor their aging process, achieving the desired flavor balance and enhancing the overall drinking experience.

Hungarian Oak

Hungarian oak barrels, made from *Quercus frainetto* (Hungarian oak) or *Quercus petraea* (sessile oak), fall somewhere between

American and French oak in terms of flavor intensity. The pore size and tightness of the grain in Hungarian oak is between that of American and French oak but closer to French oak.

Hungarian oak provides a balanced combination of vanilla, spice, and tannins, suitable for a wide range of beer styles. It offers a unique flavor profile that falls closer to French oak in terms of intensity, making it a versatile option for brewers.

- Furfural: A compound found in Hungarian oak that adds rich, toasty notes to beer, enhancing its complexity and depth.

Although less common than American and French oak, Hungarian oak is highly prized for the unique profile it brings to both wine and beer. Scientific analysis indicates that compounds specific to Hungarian oak, such as furfural and eugenol, add rich, toasty, and spicy notes, enhancing a wide range of beer styles.

Native to Eastern Europe, particularly regions such as Hungary, Romania, and the Balkans, Hungarian oak provides a unique flavor profile that adds another layer of uniqueness to the beer. Understanding these compounds allows brewers to experiment with different barrels, achieving diverse and intriguing flavor profiles.

Flavor Contributions

Tannins

Tannins are naturally occurring compounds in oak that add astringency and structure to the beer. They interact with proteins in the beer to create a smoother mouthfeel and help stabilize the beer, reducing oxidation. The presence of tannins can enhance the beer's body and complexity, contributing to a more satisfying drinking experience.

Scientific analysis shows that tannins bind with proteins to create a smoother mouthfeel and contribute to the beer's stability, reducing the risk of oxidation and extending shelf life. This interaction helps prevent the beer from developing off-flavors over time, ensuring it remains fresh and enjoyable.

Vanillin

As previously mentioned, vanillin is a key flavor compound in oak that imparts vanilla-like aromas and flavors. The presence of vanillin can add a sweet, creamy note to the beer, particularly appealing in darker, richer beer styles.

Scientific analysis reveals that the extraction of vanillin from the oak occurs gradually over time, allowing the flavors to develop and integrate seamlessly into the beer. This slow extraction process ensures that the beer acquires a balanced and harmonious flavor profile.

Lactones

Lactones contribute to the coconut and woody aromas found in oak-aged beers. American oak is particularly high in lactones, making it a popular choice for beers that aim for a pronounced woody character.

Scientific analysis indicates that lactones are extracted from the oak over time, contributing to the beer's aroma and flavor profile. The presence of lactones creates a complex and inviting bouquet, enhancing the sensory appeal of the beer.

Previous Barrel Use

The previous contents of the barrel also play a crucial role in the final flavor profile of the beer. Bourbon barrels, for instance, contribute flavors of caramel, toffee, and vanilla, while wine barrels can add fruity, tannic, and vinous notes. The residual flavors from the previous use create a synergy with the beer, resulting in a unique and complex end product.

Scientific analysis shows that the interaction between the residual flavors and the beer can significantly enhance the complexity and depth of the final product. Understanding the impact of previous barrel use allows brewers to select barrels that will complement and enhance the flavors of their beer.

Wood Treatment

Sanitizing Barrels

Sanitizing barrels is a critical step in the aging process to prevent contamination and ensure the beer remains uncontaminated. Several methods can be used to sanitize barrels effectively, each with its own advantages and considerations.

- Steaming: A method of sanitizing barrels by using high-pressure steam to kill bacteria and other microorganisms, ensuring the barrel is clean without introducing chemicals. This method preserves the integrity of the barrel while ensuring it is ready for use.
- Ozone Treatment: A modern method of barrel sanitization that uses ozone gas to kill bacteria, yeast, and mold without leaving any residue. Ozone treatment ensures the barrel is free of contaminants and is increasingly popular due to its effectiveness and minimal impact on the barrel's integrity.
- Chemicals for Barrel Sanitization: Substances such as sulfur dioxide or potassium metabisulfite are used to sanitize barrels by killing microorganisms. These chemicals must be thoroughly rinsed out before use to prevent residual flavors from affecting the beer. Chemical sanitizers offer a quick and efficient solution but require careful handling.

Wood Chips and Spirals

For brewers without access to traditional barrels, wood chips and wood spirals offer practical alternatives for imparting wood flavors to beer.

- Wood Chips: Small pieces of oak or other woods added directly to the fermenter or conditioning vessel to impart wood flavors to beer. They provide quick flavor extraction due to their high surface area. Adjusting the quantity and contact time allows brewers to control the intensity of the flavors.

- Wood Spirals: Coiled pieces of wood that provide a larger surface area for flavor extraction. They offer a controlled and gradual release of flavors, mimicking the effects of barrel aging in a more manageable format. Spirals are ideal for longer aging periods and allow for a more nuanced flavor development.

These tools are particularly useful for homebrewers or small-scale brewers looking to experiment with wood aging without committing to full barrels.

Practical Tips

Aging Time

Regularly taste the beer during the aging process to monitor the development of flavors. Over-aging can lead to excessive woodiness or harsh tannins, while under-aging may result in a lack of depth and complexity. Adjust the aging time based on the intensity of the flavors you want to achieve.

- Lighter Beers: May only need a few weeks of aging to acquire subtle wood characteristics.
- Robust, High-Alcohol Beers: Can benefit from several months of aging to develop deeper complexity.

By tasting regularly, brewers can determine the optimal aging time, ensuring the beer reaches its peak flavor.

Blending

Blending is the practice of mixing barrel-aged beer with fresh beer or combining different batches of barrel-aged beer to create a balanced and harmonious final product with enhanced complexity. This allows brewers to fine-tune the intensity of the wood and other flavors.

- Balance: Achieve the desired flavor profile by adjusting the proportions of aged and fresh beer.
- Complexity: Combining different aged beers can create unique flavor combinations.

Experimenting with blending provides flexibility and control, enabling brewers to create the perfect balance of flavors.

Rationalism, Empiricism, and the Scientific Method

Applying Rationalism

Rationalism involves using deductive reasoning to develop theories and principles for barrel aging. Understanding the chemical composition of different oak types allows brewers to predict how barrels will influence the beer's flavor.

For example:

- Predicting Flavor Profiles: Knowing that American oak is high in lactones and vanillin, a brewer can anticipate a beer aged in these barrels will have strong vanilla and coconut notes.
- Selecting Wood Types: Choosing French oak for its subtle tannins and spice notes to complement a lighter beer style.

Rational thought allows brewers to anticipate and control the effects of barrel aging, ensuring consistent and high-quality results.

Applying Empiricism

Empiricism involves gathering data through careful observation and experimentation. By experimenting with different aging times, wood types, and monitoring the resulting flavors, brewers can refine their techniques based on real-world observations.

For instance:

- Experimenting with Aging Times: Adjusting the duration of aging to see how it affects flavor development.
- Testing Different Barrels: Aging the same beer in American, French, and Hungarian oak barrels to compare flavor outcomes.

Empirical evidence helps improve brewing practices and enhances the quality of barrel-aged beers.

Using the Scientific Method

The scientific method provides a structured approach to solving brewing problems and innovating:

1. Observation: Noticing the effects of barrel aging on beer flavor.
2. Hypothesis Formation: Predicting that a particular type of barrel or aging time will influence the beer in a specific way.
3. Experimentation: Testing the hypothesis by aging beer under controlled conditions.
4. Data Analysis: Analyzing the flavors and characteristics of the aged beer.
5. Replicability: Ensuring that the results can be replicated by following the same procedures.
6. Peer Review: Sharing findings with the brewing community for feedback and validation.

This systematic approach ensures that brewing practices are continually refined and improved based on solid evidence.

Practical Applications and Techniques

Barrel Selection

- Type of Oak: Choose between American, French, or Hungarian oak based on desired flavor profiles.
- Previous Barrel Use: Consider barrels that previously held bourbon, wine, or other spirits for additional complexity.

Selecting the appropriate barrel is crucial for achieving the desired flavor characteristics.

Barrel Preparation

- Cleaning and Sanitizing: Properly clean and sanitize barrels to prevent contamination. Techniques include steaming, ozonating, or using chemical sanitizers.
- Inspection: Check for leaks or damage before use.

Thorough preparation ensures that the beer remains free of unwanted microorganisms and that the flavors develop as intended.

Barrel Maintenance

- Regular Inspection: Monitor barrels for signs of leakage or contamination.
- Hydration: Keep barrels hydrated to prevent drying out and cracking.
- Storage Conditions: Store barrels under appropriate temperature and humidity conditions.

Proper care extends the useful life of barrels and maintains the integrity of the flavors imparted to the beer.

Using Wood Chips and Spirals

For those without access to traditional barrels:

- Wood Chips: Ideal for shorter aging periods; adjust quantity and contact time to achieve desired intensity.
- Wood Spirals: Provide a controlled release of flavors suitable for longer aging periods.

These methods allow brewers to experiment with wood aging in a manageable and cost-effective way.

Modern Advancements and Innovations

Recent advancements in technology and methodology have revolutionized barrel aging. Automation and data analytics allow for precise control over aging conditions, enhancing consistency and quality.

Automation

- Definition: The use of advanced technology to control brewing processes with minimal human intervention.
- Benefits: Enhances efficiency, consistency, and quality in modern brewing operations.
- Applications: Automated systems for temperature control, humidity regulation, and monitoring of aging conditions.

Temperature-Controlled Barrels

- Function: Barrels equipped with systems to precisely control aging temperatures.
- Benefits: Allows for consistent aging conditions, reducing variability in flavor development.
- Examples: Temperature jackets or integrated cooling/heating systems.

Data Analytics

- Definition: Analyzing large datasets to optimize processes and predict trends.
- Benefits: Identifies optimal aging conditions, predicts flavor outcomes, and improves decision-making.
- Applications: Monitoring fermentation kinetics, flavor compound development, and aging progress.

These innovations help brewers achieve greater consistency and control in their barrel aging processes, leading to higher-quality beers with complex and desirable flavors.

Challenges and Solutions

Barrel aging presents several challenges, including contamination, over-extraction, and barrel maintenance. Addressing these issues involves:

Contamination Prevention

- Rigorous Cleaning and Sanitizing: Implement thorough protocols to ensure barrels are free of unwanted microorganisms.
- Proper Storage: Maintain appropriate environmental conditions to prevent microbial growth.

Managing Extraction

- Regular Monitoring: Taste the beer periodically to avoid over-extraction of tannins and wood flavors.
- Adjusting Aging Time: Modify the duration based on flavor development.

Barrel Care

- Maintenance: Repair leaks promptly, keep barrels hydrated, and store under suitable conditions.
- Rotation: Rotate barrels as needed to ensure even aging and prevent stagnation.

Proper management of these challenges ensures the barrel aging process enhances the beer without introducing negative qualities.

Narrative Example

Imagine you are a brewer working on a new barrel-aged stout. You start by selecting American oak barrels that previously held bourbon. The barrels arrive at your brewery, and you begin the preparation process by steaming them to ensure they are free of any unwanted microorganisms. The rich aroma of bourbon wafts from the barrels, hinting at the flavors they will impart to your stout.

You carefully transfer the stout into the barrels, knowing that the alchemy of aging will transform it over the coming months. Every few weeks, you sample the beer, noting the evolving flavors. The stout begins to take on hints of vanilla, coconut, and a subtle smokiness from the charred oak.

After three months, the beer reaches the perfect balance of flavors. Deciding to blend, you combine the barrel-aged stout with a fresh batch, creating a final product that has the complexity of barrel aging with the freshness and balance of newly brewed beer. The result is a rich, layered stout with a velvety mouthfeel and a symphony of flavors that dance on the palate.

By mastering barrel aging and wood treatment, you have elevated your brewing skills, creating a beer that tells a story with every sip—a testament to the artistry and science of brewing.

Conclusion

Barrel aging and wood treatment are deeply intertwined with the principles of rationalism, empiricism, and the scientific method. These approaches have guided the evolution of techniques, enabling brewers to achieve complex and desirable flavors. By combining theoretical knowledge with practical experimentation, brewers can continually refine their methods and produce exceptional barrel-aged beers.

The themes explored in this chapter are foundational for understanding how traditional methods can be refined and enhanced through scientific inquiry and practical application. The principles of continuous improvement and systematic experimentation will always be central to our exploration of the art and science of brewing.

> *Barrel aging teaches that excellence cannot be rushed. In the embrace of wood and time, beer becomes more than beverage—it becomes memory, patience, and transformation made liquid. Mastery lies not in control, but in knowing when to wait, when to blend, and when to let go.*

**

Review Questions

True/False Questions

1. True or False: Barrel aging is a technique that enhances the flavor and complexity of beer by introducing unique characteristics through the interaction of beer with the wood of the barrel.

2. True or False: American oak barrels are known for their subtle flavors and are typically used for lighter beer styles.

3. True or False: The presence of tannins in oak barrels contributes to the astringency and structure of the beer, enhancing its mouthfeel and stability.

4. True or False: Vanillin, a key flavor compound in oak, imparts sweet, creamy notes to the beer, making it especially appealing in darker beer styles.

5. True or False: Regularly tasting beer during the aging process is important to avoid over-aging, which can result in excessive woodiness or harsh tannins.

Multiple Choice Questions

6. Which type of oak barrel is known for providing a balanced combination of vanilla, spice, and tannins?
A) American Oak
B) French Oak
C) Hungarian Oak
D) Brazilian Oak

7. What is a modern and highly effective method of barrel sanitization that leaves no residue?
A) Steaming
B) Ozone treatment
C) Using sulfur dioxide
D) Using potassium metabisulfite

Brewer Vignettes

8. Brewer Vignette 1: You notice that your beer has developed a pronounced vanilla aroma and flavor. Which compound is most likely responsible, and what steps could you have taken to enhance this characteristic?
A) Tannins; adjust mash temperature.
B) Lactones; increase hop additions.
C) Vanillin; choose American oak barrels and monitor aging time.
D) Furfural; extend fermentation period.

9. Brewer Vignette 2: As a brewer, you want to create a beer with a pronounced coconut and woody aroma. What steps would you take to achieve this?
A) Use French oak barrels and add more yeast.
B) Select American oak barrels high in lactones and monitor the aging process.
C) Choose Hungarian oak barrels and increase carbonation levels.
D) Avoid using any oak barrels and focus on malt profile.

10. Brewer Vignette 3: You are preparing to age a beer in barrels and need to ensure they are properly sanitized. What method would you use, and why?
A) Boil the barrels; quick and effective.
B) Ozone treatment; kills bacteria and mold without leaving residue.
C) Add more hops to the beer; hops act as natural preservatives.
D) Avoid sanitizing; the beer's alcohol content will kill any microorganisms.

**

Correct Answers

True/False Questions

1. True
2. False (American oak barrels are known for their bold and distinctive flavors, making them suitable for robust beer styles rather than lighter ones.)
3. True
4. True
5. True

Multiple Choice Questions

6. C) Hungarian Oak
7. B) Ozone treatment

Brewer Vignettes

8. C) Vanillin; choose American oak barrels and monitor aging time.
9. B) Select American oak barrels high in lactones and monitor the aging process.
10. B) Ozone treatment; kills bacteria and mold without leaving residue.

**

Beyond The Chapter

- Bamforth, Charles W. Beer: Tap into the Art and Science of Brewing. Oxford University Press, 2009.
    - *Explores the history and techniques of barrel aging and wood treatment in brewing.*
- Fix, George. Principles of Brewing Science: A Study of Serious Brewing Issues. Brewers Publications, 1999.
    - *Discusses the scientific principles behind barrel aging and the impact on beer flavor.*
- Papazian, Charlie. The Complete Joy of Homebrewing. HarperCollins, 2014.
    - *Provides practical advice on barrel aging and wood treatment techniques for homebrewers.*

- Palmer, John J. How to Brew: Everything You Need to Know to Brew Great Beer Every Time. Brewers Publications, 2017.
    - *Offers guidance on incorporating barrel aging into the brewing process.*
- Zainasheff, Jamil. Brewing Classic Styles. Brewers Publications, 2007.
    - *Discusses techniques for barrel aging and wood treatment in classic beer styles.*
- White, Chris, and Jamil Zainasheff. Yeast: The Practical Guide to Beer Fermentation. Brewers Publications, 2010.
    - *Explores the impact of barrel aging on yeast performance and fermentation outcomes.*
- Oliver, Garrett. The Oxford Companion to Beer. Oxford University Press, 2011.
    - *Provides an encyclopedic overview of barrel aging and wood treatment in brewing.*
- Noel, Josh. Barrel-Aged Stout and Selling Out. Chicago Review Press, 2018.
    - *Examines the cultural and commercial aspects of barrel-aged beers.*
- Cantwell, Dick, and Peter Bouckaert. Wood & Beer: A Brewer's Guide. Brewers Publications, 2016.
    - *A comprehensive guide on the techniques and benefits of using wood in brewing.*

Weblinks

- American Homebrewers Association - Barrel Aging Beer: https://www.homebrewersassociation.org/how-to-brew/5-tips-barrel-aging-beers-avery-brewing-co/
- Brew Your Own - Guide to Barrel Aging: https://byo.com/?s=Barrel+Aging
- Craft Beer & Brewing - Barrel Aging Basics: https://beerandbrewing.com/search/?search=Barrel+Aging

- The Barley Blog - Barrel Aging Beer: http://www.thebarleyblog.com/index.php?s=Barrel
- Distillery Trail - Barrel Aging Science: https://www.distillerytrail.com/?s=Beer+Barrel+Aging

These references offer valuable insights into both the theoretical and practical aspects of barrel aging and wood treatment.

***

Chapter 14: The Art of Transformation—Mastering Sour Beers

Sour beers epitomize the art of transformation, where wild yeast and bacteria craft complexity, blending ancient traditions with scientific precision.

Imagine the tangy, refreshing taste of a well-crafted sour beer—a brew that both challenges and delights the palate with its complexity. Sour beers have surged in popularity due to their unique flavors and the intricate processes involved in their production. The brewing techniques for sour beers set them apart from other styles, offering a diverse range of flavors from tart and tangy to funky and earthy. These beers are often more challenging to brew because of the complex processes and the use of wild yeast and bacteria. Rationalism, empiricism, and the scientific method play crucial roles in developing and perfecting sour beer recipes. Understanding and applying these principles can help brewers achieve consistently high-quality sour beers that captivate the palate.

Sour Beers

Lambics

Lambic is a traditional Belgian beer brewed using spontaneous fermentation, where the wort is exposed to natural yeast and bacteria in the open air. This method, unique to the Senne Valley of Belgium, allows the local microflora to inoculate the wort, imparting flavors that are characteristic of the region's terroir—the unique environmental factors, including local microflora, that influence the character of a beer.

The brewing process begins with a blend of malted barley and a high proportion of unmalted wheat, providing a complex carbohydrate profile for the diverse microorganisms to ferment.

Aged hops are used not for their bitterness but for their preservative qualities, as they have lost much of their alpha acids but retain antibacterial properties. The wort is then cooled overnight in a coolship—a shallow, open fermentation vessel traditionally used in brewing, especially for spontaneously fermented beers like Lambics. The wide surface area of the coolship allows the hot wort to cool quickly while being exposed to wild yeasts and bacteria from the environment, initiating fermentation.

After fermentation begins, the beer is transferred to oak barrels, where it is aged for one to three years. The aging process allows for a slow and complex interaction between the beer, the wood, and the resident microorganisms. The oak barrels contribute tannins and subtle flavors, while the microorganisms continue to ferment complex sugars, developing layers of flavor that evolve over time.

Historically, Lambic brewing dates back centuries, with origins in the Brussels region. The unique terroir of the Senne Valley is essential to the character of Lambic beers. The centuries-old techniques, such as spontaneous fermentation and barrel aging, continue to define the modern interpretation of Lambic beers, maintaining a bridge between the past and present.

Practical Tips:

- Ingredients: Use a blend of malted barley, unmalted wheat, and aged hops.
- Fermentation: Cool the wort in a coolship overnight to capture wild yeast and bacteria from the environment.
- Aging: Age the beer in oak barrels for one to three years, during which it develops its unique profile through the slow interaction with the wood and resident microorganisms. Regular tasting throughout the aging process allows brewers to monitor the development of flavors and make adjustments

as needed, ensuring the final product is a true representation of the Lambic style.

Gose

Gose is a German-style sour beer brewed with coriander and salt, undergoing lactic fermentation to impart a refreshing tartness. Originating in Goslar, Germany, along the Gose River, this beer style is known for its light body, high carbonation, and distinctive flavor profile. The addition of salt provides a slight saline note, while coriander, a spice used in brewing to add citrusy and spicy notes, contributes a subtle complexity.

The use of Lactobacillus, a type of bacteria that produces lactic acid, is essential in Gose production. The lactic acid bacteria convert sugars, primarily glucose, into lactic acid—a process known as lactic fermentation. This imparts the beer's characteristic tartness—a sharp, tangy flavor resulting from the presence of acids like lactic acid.

Gose nearly disappeared in the mid-20th century but has seen a revival in modern craft brewing. This resurgence has brought new life to the style, with brewers experimenting with additional ingredients and variations on the traditional recipe. The adaptability of Gose to contemporary tastes showcases the timeless appeal of this unique beer.

Practical Tips:

- Ingredients: Use malted barley, wheat, coriander, and salt.
- Fermentation: Control fermentation temperatures to ensure the proper development of Lactobacillus, which imparts the desired tartness.
- Balance: Adjust the levels of coriander and salt to achieve a harmonious flavor profile that enhances the beer's refreshing character without overpowering it. This delicate balance is achieved through careful measurement and adjustments based on sensory evaluations, ensuring each batch of Gose is both unique and true to its origins.

Berliner Weisse

Berliner Weisse is a low-alcohol wheat beer known for its light, tart, and effervescent character. It undergoes lactic fermentation, resulting in a highly refreshing beer perfect for summer. The delicate sourness and high carbonation make it an ideal candidate for blending with flavored syrups.

Historically popular in 19th-century Berlin, Berliner Weisse experienced a decline before being revived by craft brewers seeking to explore historical styles. Traditionally, it is served with flavored syrups like raspberry or woodruff—a sweet, herbal syrup —to balance its tartness, offering a customizable drinking experience that caters to individual preferences.

Practical Tips:

- Fermentation: Use a combination of brewer's yeast and Lactobacillus. The brewer's yeast ferments the sugars into alcohol, while Lactobacillus produces lactic acid, contributing a clean, tangy sourness.
- Acidity: Monitor pH levels to achieve the desired tartness, and maintain proper fermentation conditions to ensure the delicate balance of flavors.
- Serving: Traditionally served with flavored syrups like raspberry or woodruff to balance the sourness and add complexity.

Flanders Red Ale

Flanders Red Ale is a Belgian-style sour beer known for its deep red color and complex sour flavors, often with fruity and malty notes. It is typically aged in oak barrels, which contribute to its depth and complexity through the slow interaction with the wood and resident microorganisms. The blending of younger and older beers is a common practice to achieve a balanced final product.

Originating in the West Flanders region of Belgium, this style has a rich history of barrel aging and blending to achieve balance and complexity. The careful management of the aging process and the art of blending are key to creating a harmonious beer.

Practical Tips:

- Aging: Age the beer in oak barrels for extended periods, allowing the flavors to develop and deepen.
- Blending: Blend older and younger beers to achieve the desired balance and complexity, creating a layered and nuanced final product.
- Flavor Development: Regular tasting and monitoring during the aging process are crucial to determine the optimal blending ratios and aging times, ensuring each batch meets the high standards of this traditional style.

Modern Approaches

Kettle Souring

Kettle souring is a method of souring the wort with Lactobacillus before boiling, allowing for faster production and greater control over the level of sourness compared to traditional methods. This technique enables brewers to produce sour beers without the extended aging times required for spontaneous fermentation or mixed fermentation methods.

Benefits: Kettle souring reduces the risk of contamination, as the wort is boiled after souring, killing the Lactobacillus and any other microorganisms before fermentation with brewer's yeast. This method also provides more control over the final acidity of the beer, allowing brewers to fine-tune their recipes with precision.

Practical Tips:

- Temperature Control: Maintain the wort temperature around 110–120°F (43–49°C) during the souring process to optimize Lactobacillus activity.
- pH Monitoring: Regularly check pH levels to achieve the desired sourness, typically aiming for a pH of 3.2 to 3.5.

- Boil: Boil the wort after souring to kill the Lactobacillus, allowing the brewer to proceed with normal fermentation using brewer's yeast.

Mixed Fermentation

Mixed fermentation involves using a combination of brewer's yeast, wild yeast, and bacteria. This method creates a diverse range of flavors and often requires extended aging. The interaction between different microorganisms can produce complex, layered beers with unique characteristics.

Flavor Development: Mixed fermentation can yield sour, funky, fruity, and earthy flavors, contributing to a complex and layered beer. The diversity of microorganisms involved adds depth and intrigue to the final product, making each beer a unique expression of the brewer's craft.

Practical Tips:

- Microbial Cultures: Select appropriate strains of yeast and bacteria based on the desired flavor profile. Common microorganisms used include Brettanomyces (wild yeast), Lactobacillus, and Pediococcus.
- Fermentation Management: Monitor the fermentation process and adjust conditions as needed to ensure the optimal development of flavors.
- Aging: Allow sufficient time for aging, as the interaction between the various microorganisms and the beer continues to evolve, resulting in a rich and multifaceted final product.

Mixed Fermentation Microorganisms

Brettanomyces ("Brett")

Brettanomyces, commonly known as "Brett," is a wild yeast known for producing unique flavors in beer, ranging from fruity and funky to earthy and barnyard-like. It can ferment complex sugars that other yeasts cannot, adding depth and complexity to sour beers.

Historically, Brettanomyces was often an unwanted contaminant, but its unique characteristics have been embraced by modern brewers. Today, it is intentionally used to create distinctive and complex beers.

Practical Tips:

- Strain Selection: Choose Brettanomyces strains that complement the desired beer style, as different strains can produce a wide range of flavors.
- Fermentation Management: Monitor fermentation closely, as Brett can be slow and unpredictable. It may require extended fermentation times.
- Flavor Development: Be patient with Brettanomyces, as it can take several months to develop its full flavor potential. The resulting beer is often a testament to the brewer's skill and understanding of this wild yeast, offering a distinctive and memorable drinking experience.

Lactobacillus

Lactobacillus is a type of bacteria that produces lactic acid, contributing a clean, tangy sourness to beer. It is commonly used in both traditional and modern sour beer production.

Practical Tips:

- Pitching Rates: Use appropriate pitching rates to ensure a healthy fermentation and efficient acid production.
- Fermentation Conditions: Maintain optimal temperatures for Lactobacillus activity, typically between 95–115°F (35–46°C).
- Sourness Levels: Monitor pH levels to ensure the beer develops the desired level of acidity, resulting in a balanced and refreshing sour beer.

Pediococcus

Pediococcus is a type of bacteria that produces intense sourness and can develop diacetyl, adding a buttery flavor. It is used in sour beer production, requiring careful management to balance its characteristics.

Practical Tips:

- Managing Diacetyl: Use extended aging or additional fermentation with Brettanomyces to reduce diacetyl levels, as Brett can metabolize diacetyl produced by Pediococcus.
- Fermentation Timelines: Be prepared for a longer fermentation process with Pediococcus, as it can be slow to develop.
- Blending Techniques: Blend different batches to balance intense sourness and complexity, achieving a harmonious final product.

Equipment, Aging, and Blending

Separate Equipment

To prevent cross-contamination between sour and non-sour beers, it is important to maintain separate equipment for each. Sour beers involve microorganisms that can be difficult to eliminate completely from equipment. Cleaning and sanitizing practices should be rigorous, with dedicated fermenters, hoses, and other tools for sour brewing. Labeling and organizing equipment helps avoid accidental cross-contamination, ensuring the integrity of both sour and non-sour batches.

Aging

Extended aging allows for the development of complex flavors. The interaction between the beer and the aging vessel, whether a barrel, stainless steel tank, or bottle, adds depth and character. Barrels can contribute oak flavors and provide an environment for microorganisms to thrive. Stainless steel tanks offer more control over environmental conditions, while bottle conditioning allows for continued fermentation and flavor development. Regular tasting and evaluation during aging help determine the optimal aging times, ensuring the beer develops its full potential.

Blending

Blending different batches can achieve balance and enhance complexity. This technique allows brewers to combine the best qualities of multiple beers into a harmonious final product. Tasting and evaluating different batches help determine the blending ratios, ensuring consistency in the final product. Blending provides a way to fine-tune the beer's flavor profile, creating a unique and satisfying drinking experience.

Rationalism, Empiricism, and the Scientific Method

Applying Rationalism

Rationalism involves using deductive reasoning to develop theories and principles in the brewing of sour beers. Understanding the principles of fermentation and microbial activity allows brewers to predict and control the production process. This theoretical knowledge guides the selection of yeast and bacteria strains and the management of fermentation conditions, ensuring consistent and high-quality results.

For example, knowing that Lactobacillus thrives at certain temperatures and produces lactic acid, a brewer can design a fermentation schedule that optimizes sourness without introducing unwanted flavors.

Applying Empiricism

Empiricism involves developing and refining sour beer recipes through careful observation, experimentation, and sensory analysis. By recording the outcomes of different brewing methods and ingredients, brewers can improve their techniques based on empirical data. This bottom-up approach allows for continuous refinement and improvement, leading to better quality and more consistent results.

Brewers might experiment with different strains of Brettanomyces or adjust aging times based on tasting notes, using empirical evidence to guide their decisions.

Using the Scientific Method

The scientific method provides a structured approach to optimizing sour beer brewing processes:

1. Observation: Observe the effects of fermentation and flavor development in sour beers.
2. Hypothesis Formation: Formulate a hypothesis about how certain microorganisms or fermentation conditions will impact the beer's flavor.
3. Experimentation: Conduct experiments by varying one variable at a time, such as using different yeast strains or adjusting fermentation temperatures.
4. Data Analysis: Analyze the results, including pH levels, flavor profiles, and fermentation performance.
5. Replicability: Repeat successful experiments to confirm results and ensure consistency.
6. Peer Review: Share findings with the brewing community for feedback and validation, refining methods based on collective knowledge.

This systematic approach ensures that brewing practices are continually refined and improved based on solid evidence.

Narrative Example

Imagine a brewer setting out to develop a new sour beer recipe. They begin by applying rationalism, using their theoretical knowledge to select microbial cultures that will produce the desired flavors—perhaps a combination of Lactobacillus for tartness and Brettanomyces for complexity.

They proceed with empiricism, carefully monitoring pH, temperature, and flavor development throughout fermentation. They take detailed notes on the fermentation progress and sensory evaluations, adjusting variables as needed.

After aging the beer, they blend different batches to achieve a balanced flavor profile, applying their sensory analysis to determine the optimal blend.

Using the scientific method, the brewer refines their process by testing hypotheses about how changes in fermentation conditions affect the final product. They share their findings with other brewers, receive feedback, and continue to improve their techniques.

This comprehensive approach highlights the importance of combining theoretical knowledge with practical experimentation to achieve exceptional results.

Conclusion

Sour beers and mixed fermentation represent a rich tradition and a modern exploration of brewing complexity. Continuous learning, experimentation, and the application of rationalism, empiricism, and the scientific method are essential for creating unique and high-quality sour beers. By embracing these principles, brewers can push the boundaries of flavor and innovation in their craft.

The journey of mastering sour beer production is a testament to the enduring blend of art and science in brewing, offering endless opportunities for creativity and discovery.

> *To brew sour beer is to practice patience in the face of uncertainty—to embrace wildness not as a flaw, but as a teacher. True mastery lies in guiding, not commanding; in listening, not insisting. Sourness, like truth, reveals itself slowly, and only to those who let go of control.*

**

Review Questions

True/False Questions

1. True or False: Sour beers are produced using techniques that often involve wild yeast and bacteria, making the brewing process more challenging compared to conventional beer styles.

2. True or False: Lambic beers are traditionally brewed in the Senne Valley of Belgium and utilize spontaneous fermentation.

3. True or False: Gose is a German-style sour beer that includes coriander and salt in its recipe.

4. True or False: Kettle souring is a traditional method that requires extended aging periods.

5. True or False: Brettanomyces, a wild yeast used in mixed fermentation, can produce fruity, funky, and earthy flavors.

Multiple Choice Questions

6. Which of the following is a primary characteristic of Berliner Weisse?
A) High alcohol content
B) Low carbonation
C) Tart and effervescent character
D) Strong bitterness

7. What is a practical tip for preventing cross-contamination when brewing sour and non-sour beers?
A) Use the same equipment for both sour and non-sour beers.
B) Maintain separate equipment for sour and non-sour beers.
C) Only use chemical sanitizers.
D) Avoid using wild yeast and bacteria.

Brewer Vignettes

8. Brewer Vignette 1: You are brewing a traditional Gose and want to ensure the correct balance of flavors. What steps would you take to achieve this?
A) Ignore the levels of coriander and salt.
B) Use aged hops for bitterness.
C) Adjust the levels of coriander and salt to balance the tartness.
D) Add more yeast for higher alcohol content.

9. Brewer Vignette 2: As a brewer, you want to experiment with kettle souring for a quicker production of sour beer. What is a crucial step in this process?
A) Avoid monitoring pH levels.
B) Boil the wort before souring.
C) Maintain the wort at around 110-120°F (43-49°C) during souring.
D) Use high-temperature fermentation.

10. Brewer Vignette 3: You are preparing to blend different batches of Flanders Red Ale to achieve a balanced final product. What approach should you take?
A) Avoid tasting the beer during the aging process.
B) Blend older and younger beers to balance flavors.
C) Use only one batch for simplicity.
D) Ignore the flavor development in different barrels.

**

Correct Answers
True/False Questions
1. True
2. True
3. True
4. False (Correct Answer: Kettle souring is a modern method that allows for faster production of sour beers without the need for extended aging periods.)

5. True

Multiple Choice Questions

6. C) Tart and effervescent character

7. B) Maintain separate equipment for sour and non-sour beers.

Brewer Vignettes

8. C) Adjust the levels of coriander and salt to balance the tartness.

9. C) Maintain the wort at around 110-120°F (43-49°C) during souring.

10. B) Blend older and younger beers to balance flavors.

**

Beyond The Chapter

- Oliver, Garrett. The Oxford Companion to Beer. Oxford University Press, 2011.
  - *Provides an overview of sour beers and mixed fermentation techniques.*
- Sparrow, Jeff. Wild Brews: Beer Beyond the Influence of Brewer's Yeast. Brewers Publications, 2005.
  - *Explores the production of sour beers and the use of wild yeast strains in brewing.*
- Brewers Association and American Homebrewers Association websites for articles, forums, and educational resources.
  - *Offers additional resources and discussions on sour beers and mixed fermentation.*

Weblinks

- American Homebrewers Association - Brewing Sour Beers: https://www.homebrewersassociation.org/?s=sour+beer
- Craft Beer & Brewing - Guide to Brewing Sour Beers: https://beerandbrewing.com/search/?search=sour+beer
- Serious Eats - The Science of Sour Beer: https://www.seriouseats.com/search?q=sour+beer

- Brulosophy - Sour Beer Experiments: https://brulosophy.com/?s=Sour+Beers

These references provide valuable insights into traditional and modern sour beer brewing techniques, helping brewers deepen their understanding and improve their practices.

***

Chapter 15: Brewing Sustainability—Blending Tradition with Innovation

Sustainable brewing blends tradition with innovation, proving that responsible stewardship and exceptional beer go hand in hand.

Imagine a brewery where every aspect—from water usage to energy consumption—is meticulously managed for minimal environmental impact. Sustainability in brewing involves adopting practices aimed at reducing the environmental impact of brewing, such as efficient water management, renewable energy use, and sustainable ingredient sourcing. Given the significant resource consumption and waste production in brewing, implementing sustainable practices ensures the long-term viability of the industry and helps mitigate environmental harm.

Rationalism, empiricism, and the scientific method play vital roles in developing and implementing these practices effectively. By integrating these approaches, brewers can enhance their processes, reduce their environmental footprint, and contribute positively to the global effort to preserve natural resources.

Environmental Impact

Water as a Key Ingredient

Water is essential in every stage of the brewing process—from mashing and boiling to cleaning, cooling, and packaging. Breweries, especially large-scale operations, have a significant water footprint, often using several gallons of water for every gallon of beer produced. Water management—the efficient use and recycling of water in brewing operations—is critical not only for sustainability but also for cost reduction and operational efficiency.

Historically, breweries have relied heavily on local water sources, with early brewers often unaware of the long-term impact of their water consumption. Over time, traditional methods have evolved to incorporate more efficient water usage practices. Modern breweries are adopting advanced technologies and strategies to minimize water usage and promote sustainability.

Water Recycling

One of the key components of sustainable brewing is water recycling—the process of treating and reusing water within the brewery to reduce overall water consumption. Advanced systems like membrane bioreactors help breweries recycle process water for non-potable uses, such as cleaning, cooling, and irrigation. Membrane bioreactors are advanced water recycling systems that treat and reuse water within the brewery, significantly reducing the demand for fresh water.

Practical Tips:

- Implementation: Install water recycling systems like membrane bioreactors. Consider the costs of equipment and maintenance while evaluating the return on investment (ROI) to justify the initial expenditure. Work with professionals to design a system tailored to your brewery's needs.
- Maintenance: Regularly monitor and maintain the equipment to ensure optimal performance. Schedule routine inspections and servicing to prevent breakdowns and extend the lifespan of the system.

Rainwater Harvesting

Another sustainable practice is rainwater harvesting—collecting and using rainwater for non-potable purposes in the brewing process to reduce freshwater consumption. By capturing rainwater, breweries can reduce their reliance on municipal water supplies and make use of a free and abundant resource.

Practical Tips:

- Setup: Install collection systems, such as rooftop gutters, storage tanks, and filtration systems, to collect and store rainwater. Ensure compliance with local regulations and design the system to prevent contamination.

- Usage: Use harvested rainwater for non-potable applications like cleaning, cooling, or irrigation. Treat the water as necessary to ensure it is suitable for its intended use.

Efficient Brewing Equipment

Modern, water-efficient brewing equipment can significantly reduce water usage. Technologies such as dry cooling systems and high-efficiency cleaning systems are designed to minimize water consumption without compromising performance.

Practical Tips:

- Equipment Selection: Invest in water-efficient equipment. Look for certifications and reviews that highlight water-saving features. Consider technologies like clean-in-place (CIP) systems that use less water and chemicals for cleaning brewing equipment.
- Maintenance: Ensure regular maintenance to keep equipment operating efficiently. Conduct routine checks and servicing to prevent leaks, breakdowns, and inefficiencies.

Renewable Energy Sources

Energy consumption is another critical aspect of sustainability in brewing. Breweries use significant amounts of energy for heating, cooling, and operating equipment. Incorporating renewable energy sources—energy from naturally replenishing resources like solar and wind power—helps reduce greenhouse gas emissions and dependence on fossil fuels.

Solar Panels

Solar panels are devices that convert sunlight into electricity, which can be used to power brewing operations or heat water in the brewing process. Installing solar panels provides a renewable energy source that can significantly reduce a brewery's carbon footprint and operational costs over time.

Practical Tips:

- Installation: Select appropriate sites for solar panel installation, such as rooftops or adjacent land with good sun exposure. Size the system correctly based on energy needs. Work with professional installers to ensure compliance with regulations and optimal performance.
- Maintenance: Regularly clean and inspect the panels to maintain maximum efficiency. Keep them free of dust, debris, and shading that can reduce energy output.

Wind Turbines

In locations with suitable wind conditions, wind turbines can generate renewable energy for breweries. These devices convert wind energy into electricity, providing an additional or alternative renewable energy source.

Practical Tips:

- Feasibility: Conduct a site analysis to determine if wind energy generation is viable. Consider average wind speeds, zoning regulations, and potential environmental impacts.
- Permits and Compliance: Secure necessary permits and ensure compliance with local regulations. Engage with the community to address any concerns related to noise or visual impact.

Energy-Efficient Brewing Systems

Advances in energy-efficient brewing systems—equipment and technologies designed to minimize energy consumption—help reduce a brewery's carbon footprint and operational costs. Examples include:

- Heat Recovery Systems: Technologies that capture and reuse energy that would otherwise be wasted. For instance, capturing heat from boiling wort to preheat incoming water, thereby reducing energy needed for heating.

- Energy-Efficient Boilers and Chillers: Equipment designed to operate with less energy while maintaining performance.

Practical Tips:

- Technology Selection: Choose energy-efficient systems based on the brewery's size and energy needs. Look for equipment with high-efficiency ratings and proven records of energy savings.
- Optimization: Regularly optimize these systems for peak efficiency. Conduct energy audits to identify and address inefficiencies, and implement control systems that adjust operations based on real-time demand.

Sustainable Ingredients

Benefits of Using Locally Sourced Ingredients

Sourcing ingredients locally supports local economies, reduces transportation-related emissions, and ensures fresher products. Locally sourced ingredients—obtained from local farms and suppliers—help promote sustainability and create unique beer profiles that reflect the local terroir.

Practical Tips:

- Building Relationships: Establish relationships with local farmers and suppliers. This secures a consistent and sustainable supply of ingredients and fosters a sense of community.
- Ingredient Selection: Choose high-quality local ingredients that meet brewing standards. Experiment with local varieties of hops, barley, and other adjuncts to create distinctive flavors.
- Integration: Integrate local ingredients into brewing operations. Highlight the use of local ingredients in marketing to appeal to consumers who value sustainability and local products.

Organic Brewing Advantages

Organic brewing involves using ingredients grown without synthetic pesticides, herbicides, and fertilizers. Organic practices reduce chemical runoff, promote soil health, and offer environmental and health benefits. Organic ingredients can also appeal to health-conscious consumers and those who prioritize sustainability.

Practical Tips:

- Sourcing: Source certified organic ingredients. Navigate certification processes to ensure compliance with organic standards set by regulatory bodies.
- Marketing: Market organic beers to highlight their environmental benefits. Educate consumers about the advantages of organic ingredients to attract a niche market that values sustainability.
- Challenges: Be aware of challenges such as higher costs, limited availability, and certification requirements. Overcome these challenges by:
  - Cost Management: Form cooperatives with other brewers to bulk purchase organic ingredients, reducing costs through economies of scale.
  - Long-Term Contracts: Develop long-term contracts with organic farmers to ensure a consistent supply of ingredients and support sustainable farming practices.

Waste Management

Effective waste management is critical for sustainability. Breweries produce various waste materials, including spent grain, hops, yeast, and packaging materials. Implementing strategies to reduce, reuse, and recycle waste helps minimize environmental impact.

Spent Grain

Spent grain is the leftover malt and grain from the brewing process. Rather than discarding it, spent grain can be composted—

the process of recycling organic waste into nutrient-rich soil—or used as animal feed, turning a waste product into a valuable resource. Composting spent grain reduces waste and supports local agriculture by providing valuable resources for farming.

Practical Tips:

- Composting or Animal Feed: Establish a composting system or collaborate with local farmers to use spent grain as animal feed. Spent grain is rich in nutrients and can be a valuable addition to livestock diets or soil enrichment.
- Storage: Ensure proper storage and handling of spent grain to maintain quality and prevent spoilage. Use sealed containers and timely pick-up arrangements.

Recycling

Recycling is the process of converting waste materials into new materials and objects. Recycling packaging materials like glass, aluminum, and cardboard reduces waste and conserves resources. Implementing a robust recycling program can significantly lower a brewery's environmental impact.

Practical Tips:

- Program Setup: Set up a recycling program with designated bins for sorting materials. Educate staff on proper recycling practices.
- Partnership: Partner with recycling facilities to ensure materials are processed correctly. Consider working with suppliers who use recyclable or reusable packaging.

Reducing Single-Use Plastics

Reducing single-use plastics—plastics designed to be used once and then discarded, such as straws and cups—is essential for decreasing environmental pollution and waste. Switching to biodegradable materials—those that can be broken down by natural processes into harmless substances—or reusable alternatives helps minimize the environmental impact.

Practical Tips:

- Alternatives: Use biodegradable cups, straws, and packaging materials. Consider reusable options like glassware for on-site consumption.
- Implementation: Implement changes across operations, from production to sales. Encourage customers to participate by offering incentives for using reusable containers.

Community Engagement

Community engagement involves efforts to involve and educate the local community about sustainable practices. Engaging the community fosters support, raises awareness, and can enhance a brewery's reputation.

Community Education

Educating the public about sustainable brewing practices through workshops, tours, and events raises awareness and fosters community support. These activities can also attract visitors and potential customers.

Practical Tips:

- Organization: Collaborate with local schools, community centers, and environmental organizations to organize educational events. Offer brewery tours that highlight sustainability efforts.
- Promotion: Promote events through local media, social networks, and community boards to engage a wider audience. Share stories about sustainability initiatives to inspire others.

Community Partnerships

Partnering with local farms for ingredient sourcing and waste disposal creates mutual benefits. Such partnerships can strengthen the local food system and reduce transportation emissions.

Practical Tips:

- Establishing Partnerships: Negotiate contracts and logistics with local farms to ensure a steady supply of ingredients and efficient waste disposal.
- Joint Marketing: Market joint efforts to highlight sustainability. Co-brand products or events to attract customers who value local and sustainable practices.

Supporting Environmental Initiatives

Supporting environmental initiatives, such as sponsoring or participating in local events like tree planting or river clean-up, strengthens community ties and enhances a brewery's reputation as a responsible business.

Practical Tips:

- Identification: Identify relevant initiatives that align with the brewery's values and capabilities.
- Participation: Organize volunteer recruitment among staff and customers. Provide resources or fundraising support to maximize the impact of participation.

Applying Philosophical Approaches

Rationalism—Deductive Reasoning

Rationalism helps brewers develop and implement sustainable practices by understanding and applying principles of resource management and environmental science. This top-down approach provides a theoretical foundation for sustainability initiatives, ensuring they are grounded in sound scientific principles.

For example, understanding the environmental impact of energy consumption can lead to the adoption of renewable energy sources like solar panels or wind turbines.

Empiricism—Inductive Reasoning

Empiricism refines sustainable brewing practices through observation, experimentation, and data analysis. This bottom-up approach allows brewers to develop practical solutions based on real-world data and experiences, leading to continuous improvement.

For instance, tracking water usage and identifying areas of waste can inform strategies to reduce consumption, such as implementing water recycling systems.

The Scientific Method—Structured Approach

The scientific method provides a structured approach to optimizing sustainability in brewing. The six elements include:

1. Observation: Recognize the environmental impact of brewing activities, such as high water or energy consumption.
2. Hypothesis Formation: Develop hypotheses about potential conservation techniques that could address these impacts.
3. Experimentation: Test the hypotheses by implementing conservation methods on a small scale.
4. Data Analysis: Analyze experimental data to determine the effectiveness of the techniques.
5. Replicability: Ensure that the results can be replicated in other parts of the brewery or by other brewers.
6. Peer Review: Share findings with the brewing community for feedback and validation, contributing to collective knowledge and best practices.

Narrative Example

Imagine a brewery that aims to reduce its environmental impact. The brewers begin by observing their high water and energy consumption. Applying rationalism, they understand that implementing water recycling and renewable energy could address these issues.

They form hypotheses about specific conservation methods, such as installing a membrane bioreactor for water recycling and solar panels for energy generation. Through experimentation, they implement these solutions on a pilot scale, closely monitoring performance and costs.

Analyzing the data, they find significant reductions in water usage and energy costs. The results are replicable, and they expand the systems to full-scale operations. Sharing their success with the community, they receive positive feedback and inspire others to adopt similar practices.

Additionally, they engage with local farms to source ingredients and manage waste, further enhancing sustainability. By applying rationalism, empiricism, and the scientific method, they achieve substantial reductions in their environmental footprint, setting an example for others in the industry.

Conclusion

Sustainable practices in brewing are essential for reducing environmental impact and ensuring the industry's long-term viability. Rationalism, empiricism, and the scientific method play crucial roles in achieving sustainability goals. Continuous learning, experimentation, and community involvement are key to advancing sustainable brewing practices.

By embracing these principles, brewers can make meaningful contributions to environmental conservation while continuing to produce high-quality beer. The journey towards sustainability is not only a testament to responsible stewardship of natural resources but also an ongoing commitment to innovation and excellence in the craft of brewing.

> *To brew sustainably is to live as though the world were not ours to consume, but to care for. It is the moral art of restraint in service of abundance—the realization that the best beer honors not only the palate, but the planet. In reducing waste, we enrich; in conserving, we create.*

**

Review Questions

True/False Questions

1. True or False: Implementing sustainable practices in brewing helps ensure the long-term viability of the industry and mitigate environmental harm.

2. True or False: Breweries typically use more water for cleaning and packaging than for the actual brewing process.

3. True or False: Solar panels can be used in breweries to generate electricity and reduce their carbon footprint.

4. True or False: Sourcing ingredients locally has no significant impact on the brewery's sustainability efforts.

5. True or False: Using spent grain as animal feed or for composting can help reduce brewery waste.

Multiple Choice Questions

6. What is one benefit of using advanced water recycling systems in breweries?
A) Increased water consumption
B) Reduced overall water usage
C) Higher energy costs
D) Decreased beer quality

7. Which practice can breweries adopt to minimize their use of single-use plastics?
A) Use more plastic packaging
B) Avoid recycling programs
C) Switch to biodegradable or reusable alternatives
D) Increase the use of plastic straws

Brewer Vignettes

8. Brewer Vignette 1: You are a brewer aiming to implement water recycling in your brewery. What steps would you take to ensure this system is effective and sustainable?
A) Use the recycled water only for drinking purposes.
B) Use the recycled water for brewing the beer.
C) Install water recycling systems and regularly monitor their efficiency.
D) Ignore maintenance and let the system run on its own.

9. Brewer Vignette 2: As a brewer committed to sustainability, you want to reduce your brewery's energy consumption. What approach would you take to achieve this?
A) Avoid integrating renewable energy sources like solar panels.
B) Invest in energy-efficient brewing systems and optimize them regularly.
C) Ignore energy audits and continue current practices.
D) Use older, less efficient equipment to save money.

10. Brewer Vignette 3: You are looking to source your brewing ingredients locally to support sustainability. What should you consider when establishing relationships with local farmers?
A) Develop long-term contracts with farmers to ensure a steady supply and support sustainable farming practices.
B) Avoid discussing sustainability practices with your suppliers.
C) Import ingredients from distant regions to get the best price.
D) Choose the cheapest possible ingredients without regard to quality.

**

Correct Answers
True/False Questions
1. True
2. True
3. True

4. False (Correct Answer: Sourcing ingredients locally supports local economies, reduces transportation-related emissions, and ensures fresher ingredients.)
5. True

Brewer Vignettes

6. B) Reduced overall water usage
7. C) Switch to biodegradable or reusable alternatives

Multiple Choice Questions

8. C) Install water recycling systems and regularly monitor their efficiency.
9. B) Invest in energy-efficient brewing systems and optimize them regularly.
10. A) Develop long-term contracts with farmers to ensure a steady supply and support sustainable farming practices.

**

Beyond The Chapter

- Loftus, Amelia Slayton. Sustainable Homebrewing: An All-Organic Approach to Crafting Great Beer. Storey Publishing, 2014.
  - *Discusses sustainable and organic practices in homebrewing.*
- Watson, Eric. Brewing Ecologically: Recipes for the Perfect Pint. Greystone Books, 2011.
  - *Provides recipes and techniques for environmentally friendly brewing.*
- Brewers Association and American Homebrewers Association websites for articles, forums, and educational resources.
  - *Offers resources and discussions on sustainable brewing practices.*

Weblinks

- Brewers Association - Sustainability: https://www.brewersassociation.org/sustainability/

- Sierra Nevada - Sustainability: https://sierranevada.com/sustainability/
- Brewery Resource Efficiency Guide: https://www.brewersassociation.org/?s=Sustainable

These references offer valuable insights into sustainable brewing practices, helping brewers deepen their understanding and improve their operations.

***

Chapter 16: Brewing the Future—Merging Tradition with Innovation

Future trends in brewing merge tradition with innovation, showing that the evolution of craft is a journey where the past informs the future, and creativity knows no bounds.

Imagine the future of brewing, where cutting-edge technologies and innovative techniques shape the next generation of beers. The brewing industry is continuously evolving, driven by technological advancements, shifting consumer preferences, and inventive brewing methods. Understanding these trends is crucial for staying competitive in the market. Rationalism, empiricism, and the scientific method play vital roles in predicting and adapting to these changes, ensuring that breweries remain at the forefront of the industry. As the brewing landscape shifts, those who embrace and anticipate future trends will be well-positioned to lead the industry into new frontiers of flavor, sustainability, and efficiency.

Technological Innovations

Technological innovations are fundamentally transforming the brewing process. Advances in brewing automation, data analytics, and equipment design are revolutionizing how beer is produced, enhancing quality, consistency, and efficiency.

Brewing Automation

Automated brewing systems utilize advanced sensors and software to monitor and control various aspects of the brewing process, such as temperature, pH levels, and fermentation rates. These systems offer improved consistency and quality by reducing the margin for human error and ensuring precise adherence to brewing parameters. Automation allows brewers to maintain optimal conditions throughout the brewing process, leading to a higher-quality product.

Historically, automation in brewing has evolved from manual processes to sophisticated, computer-controlled systems. Early automated systems focused primarily on basic temperature control, while modern systems provide comprehensive process management, including ingredient dosing, scheduling, and quality control. These advancements have allowed for more complex and consistent brewing operations.

Practical Tips:

- Selection: Choose an automated system that suits your brewery's size and production needs. Consider scalability—the ability of a brewing system or process to be expanded to accommodate increased production demands without compromising quality. This ensures the system can grow with your business.
- Implementation: Ensure proper infrastructure, such as reliable power supply and network connectivity, to support the automated systems. Integration with existing systems should be seamless to avoid disruptions.
- Training: Provide staff training to operate and maintain these systems effectively, ensuring all team members are proficient in using the technology. Ongoing education keeps the team updated on system capabilities.

Improved Efficiency

Automation streamlines operations, reduces labor costs, and increases productivity. Examples include automated cleaning systems that save time and water, and automated packaging lines that speed up the bottling or canning process. These systems enable brewers to scale production more efficiently, meeting increasing demand without compromising quality.

- Modular Systems: Invest in modular systems—brewing systems that can be expanded or reconfigured as needed, providing flexibility and scalability for breweries. Modular designs support growth and adaptation to changing market conditions. Implementing modular systems supports growth while maintaining efficiency.

Important Considerations:

- ROI Evaluation: Calculate the return on investment (ROI) for automated systems to justify the initial expenditures. This assessment is essential for making informed financial decisions.
- Workflow Optimization: Analyze and optimize workflows to maximize the benefits of automation. Streamlining processes can significantly enhance efficiency.
- Maintenance: Regular maintenance of equipment prevents breakdowns and ensures efficiency. Proactive maintenance schedules keep systems running smoothly and extend their lifespan.

Data Analytics

Data analytics is another transformative tool in modern brewing. It helps brewers identify trends and patterns affecting beer quality and consistency. Types of data include fermentation temperatures, yeast activity, ingredient quality, and energy consumption. By analyzing this data, brewers can make informed decisions to enhance their processes and products.

Historically, traditional record-keeping has evolved into modern data-driven decision-making. Early brewers relied on logs and observations, while modern brewers use sophisticated data analytics tools to gain deeper insights into their brewing operations.

Practical Tips:

- Implementation: Implement data analytics tools to collect and analyze brewing data. Choose software that integrates seamlessly with existing brewing systems to ensure compatibility and ease of use.

- Interpretation: Learn to interpret data and use insights to improve brewing processes. Training in data analysis can lead to significant improvements in quality and efficiency.

Improving Processes

Data analytics can reveal inefficiencies in brewing processes, such as energy consumption or ingredient usage, helping breweries optimize operations and reduce costs. Analyzing data trends allows brewers to identify areas for improvement and implement changes that enhance overall efficiency.

- Software Selection: Choose data analytics software that integrates well with existing systems. Look for features like real-time monitoring, customizable dashboards, and predictive analytics.
- Integration: Incorporate data insights into daily operations to promote continuous improvement. Regularly review data trends to stay ahead of potential issues.

Enhancing Product Quality

Detailed data on fermentation variables helps maintain optimal conditions for each batch, resulting in consistently high-quality beer. By closely monitoring key performance indicators (KPIs), brewers can ensure that every batch meets their quality standards.

- Monitoring Systems: Set up data monitoring systems to track KPIs. Automated alerts can help identify issues in real-time.
- Regular Reviews: Conduct regular data reviews to ensure consistent quality and address any issues promptly. Continuous monitoring and adjustment keep the brewing process optimal.

Emerging Styles

The brewing industry is witnessing the rise of hybrid beers—beers that combine characteristics of two different styles, offering novel flavor experiences. Examples include IPA-lagers and sour stouts, which showcase the innovative spirit of modern brewing.

IPA-Lager Hybrids

An IPA-lager is a type of hybrid beer that combines the bold hop flavors of an India Pale Ale (IPA) with the clean, crisp profile of a lager. This style appeals to consumers seeking innovative and refreshing beer options.

- Brewing Techniques: Experimenting with different hop varieties and lagering techniques helps refine recipes. The brewing process requires careful balance to achieve the desired flavor profile.
- Quality Ingredients: Choosing high-quality ingredients enhances the overall flavor. Premium hops contribute to bold flavors, while quality malts provide a smooth base.
- Monitoring Fermentation: Monitor fermentation closely to ensure the desired balance of flavors. Adjustments can be made as needed to achieve the perfect blend.

Sour Stouts

Sour stouts are a hybrid beer style that combines the rich, roasted flavors of stouts with the tartness of sour beers. This creates a complex and intriguing flavor profile.

- Souring Process: Manage the souring process carefully to avoid overpowering the stout flavors. Use controlled souring techniques, such as kettle souring or adding specific bacteria like Lactobacillus.
- Flavor Balance: Regular tastings are crucial for fine-tuning the flavor profile. Balance the sour and roasted elements to create a harmonious taste.
- Sanitation: Proper sanitation practices are vital to prevent cross-contamination between sour and non-sour beers. Use dedicated equipment for souring processes.

Historically, blending beer styles has led to the creation of modern hybrids, reflecting consumer demand for innovative flavors. The popularity of hybrid styles showcases the dynamic nature of the brewing industry.

Non-Alcoholic and Low-Alcohol Beers

The demand for non-alcoholic beers and low-alcohol beers is growing, catering to health-conscious consumers looking for lighter drinking options.

Modern Techniques

Modern techniques, such as vacuum distillation and reverse osmosis (RO), remove alcohol from beer without sacrificing flavor.

- Vacuum Distillation: A technique used to remove alcohol from beer by lowering the boiling point of alcohol through a vacuum. This process preserves the beer's flavor while making it non-alcoholic.
- Reverse Osmosis: A process where beer is passed through a semi-permeable membrane under pressure, separating alcohol and water from flavor compounds. The flavor-rich concentrate is then blended with water to create a non-alcoholic beer that retains the original taste.

These advancements have made it possible to produce flavorful options that cater to consumers seeking non-alcoholic or low-alcohol beverages.

Practical Tips:

- Technique Selection: Choose appropriate brewing techniques based on desired outcomes. Consider equipment availability and product goals.
- Recipe Adjustment: Adjust recipes to enhance flavor profiles, focusing on ingredients that contribute to a rich taste. Specialty malts and hops can add depth.
- Marketing: Market these beers to health-conscious consumers, highlighting benefits like lower calorie content and responsible drinking.

Enhancing Flavors

Brewers are creating full-flavored non-alcoholic and low-alcohol beers with bold hop profiles, rich malt characters, and innovative ingredients.

- Ingredient Experimentation: Experiment with new ingredients to create unique flavor profiles. Non-traditional ingredients, such as fruits or spices, can add exciting dimensions.
- Sensory Evaluations: Conduct sensory evaluations to ensure high-quality taste. Regular taste tests help maintain consistency and appeal.

Life-Long Learning

Staying informed and continuously learning are essential for adapting to future trends in brewing.

Attending Brewing Conferences

Attending brewing conferences and trade shows offers opportunities to learn about the latest technologies, trends, and best practices. These events provide valuable networking opportunities and access to expert knowledge.

Practical Tips:

- Conference Selection: Choose relevant conferences and events that align with your brewery's goals. Prioritize those known for high-quality content and networking opportunities.
- Maximize Opportunities: Network with other industry professionals and attend workshops. Engage with speakers and participate in discussions to gain insights.

Reading Industry Publications

Subscribing to reputable industry publications offers valuable insights into new techniques, market trends, and regulatory changes.

Practical Tips:

- Application: Apply insights from these sources to your brewing practices. Regular reviews and discussions can enhance team knowledge.

Networking with Other Brewers

Building relationships with other brewers facilitates knowledge-sharing and collaboration, leading to new ideas, joint projects, and a stronger brewing community.

Practical Tips:

- Engagement: Join local brewing associations, participate in online forums, and attend industry events.
- Collaboration: Explore collaborative brewing projects to create innovative products and share success.

Experimenting with Different Ingredients

Trying new hops, malts, yeasts, and adjuncts can lead to unique flavors. Incorporating seasonal ingredients and locally sourced produce creates distinctive brews.

Practical Tips:

- Sourcing Ingredients: Find reliable sources for new ingredients. Build relationships with suppliers.
- Test Batches: Conduct test batches to evaluate new ingredients.
- Documentation: Maintain detailed records of experiments for future reference.

Exploring New Brewing Methods

Experimenting with different brewing methods, such as barrel aging, mixed fermentation, or hybrid styles, expands brewing capabilities.

Practical Tips:

- Planning Experiments: Plan and execute experiments methodically, setting clear objectives.

- Documentation: Document results to guide future brewing decisions.

Embracing Technology

Leveraging advanced brewing technologies and data analytics enhances efficiency and quality.

Practical Tips:

- Technology Selection: Choose appropriate technology based on your brewery's needs.
- Implementation: Integrate new technologies into existing workflows.

Critical Thinking and Philosophical Approaches

Rationalism, empiricism, and the scientific method guide the development and implementation of future trends in brewing.

Rationalism—Top-Down Approach

Rationalism helps brewers understand and apply principles of technology, market analysis, and sustainability, guiding strategic decisions and long-term planning.

Empiricism—Bottom-Up Approach

Empiricism refines brewing practices through observation, experimentation, and data analysis. Continuous learning from real-world experiences drives improvement and innovation.

The Scientific Method—Structured Approach

The scientific method optimizes brewing practices, ensuring innovations are thoroughly tested and validated. Steps include:

1. Observation: Identify trends or issues in brewing processes or consumer preferences.
2. Hypothesis Formation: Develop theories about potential solutions or improvements.
3. Experimentation: Test hypotheses through controlled experiments.

4. Data Analysis: Analyze results to determine effectiveness.
5. Replicability: Ensure results can be replicated.
6. Peer Review: Share findings for feedback and validation.

Narrative Example

Imagine a brewery adopting future trends by implementing automated systems, using data analytics, and experimenting with new beer styles. They set up sensors and software to monitor brewing processes, analyzing data to identify areas for improvement.

They create an IPA-lager hybrid, combining the bold hops of an IPA with the crisp finish of a lager, appealing to customers seeking innovative flavors. Simultaneously, they develop a line of non-alcoholic beers using reverse osmosis, catering to health-conscious consumers.

Regular peer reviews and community engagement ensure continuous improvement. By sharing findings and collaborating with other brewers, they refine processes and products.

Through these efforts, the brewery stays at the industry's cutting edge, producing innovative and high-quality beers that resonate with consumers.

Conclusion

Staying informed about future trends in brewing is essential for maintaining competitiveness and innovation. Rationalism, empiricism, and the scientific method are crucial in predicting and adapting to these trends. Continuous learning, experimentation, and embracing new technologies and styles ensure success and sustainability in the brewing industry.

By staying ahead of the curve, brewers can meet changing consumer demands and contribute to the evolution of the craft. The journey towards innovation and excellence in brewing is a testament to the dynamic and ever-evolving nature of the industry. Embracing these principles ensures brewers are well-equipped to navigate the future, creating exceptional beers that set new standards.

> *To brew the future is not to abandon the past but to distill its wisdom into new forms. Innovation without memory is rootless; tradition without curiosity is stagnant. Let the brewer be both historian and explorer—crafting tomorrow's beer with yesterday's soul and today's mind.*

**

REVIEW QUESTIONS

True/False Questions

1. True or False: Automated brewing systems use advanced sensors and software to monitor and control various aspects of the brewing process.

2. True or False: Data analytics can help brewers identify trends and inefficiencies in their brewing processes.

3. True or False: IPA-lager hybrids combine the bitter hop flavors of IPAs with the crisp, clean profile of lagers.

4. True or False: Non-alcoholic beers cannot be produced with full-flavored profiles due to the lack of alcohol.

5. True or False: Attending brewing conferences and reading industry publications are effective ways for brewers to stay updated on future trends.

Multiple Choice Questions

6. What is one of the primary benefits of automated brewing systems?
A) Increased manual labor
B) Reduced consistency
C) Improved efficiency
D) Higher energy consumption

7. Which of the following methods can breweries use to create non-alcoholic and low-alcohol beers without sacrificing flavor?
A) Increasing fermentation temperature
B) Vacuum distillation and reverse osmosis
C) Using traditional fermentation methods
D) Adding more hops

Brewer Vignettes

8. Brewer Vignette 1: You are implementing automated brewing systems in your brewery. What steps should you take to ensure success?
A) Install the system without training staff.
B) Choose an automated system that suits your brewery's size and production needs.
C) Ignore the maintenance of the system.
D) Avoid integrating the system with existing workflows.

9. Brewer Vignette 2: As a brewer interested in using data analytics to improve your processes, what should you do?
A) Ignore the data and rely on intuition.
B) Use data analytics tools to collect and analyze brewing data.
C) Disregard the importance of interpreting data trends.
D) Avoid integrating data insights into daily operations.

10. Brewer Vignette 3: You want to create a new IPA-lager hybrid. What approach would you take to achieve this?
A) Ignore recipe development and use standard ingredients.
B) Experiment with different hop varieties and lagering techniques.
C) Avoid monitoring fermentation closely.
D) Use low-quality ingredients to save costs.

**

Correct Answers
True/False Questions
1. True
2. True
3. True
4. False (Correct Answer: Modern techniques, such as vacuum distillation and reverse osmosis, remove alcohol from beer without sacrificing flavor.)
5. True

Multiple Choice Questions
6. C) Improved efficiency
7. B) Vacuum distillation and reverse osmosis

Brewer Vignettes
8. B) Choose an automated system that suits your brewery's size and production needs.
9. B) Use data analytics tools to collect and analyze brewing data.
10. B) Experiment with different hop varieties and lagering techniques.

**

Beyond The Chapter

- Calagione, Sam. Brewing Up a Business: Adventures in Beer from the Founder of Dogfish Head Craft Brewery. John Wiley & Sons, 2011.
  - *Explores future trends in brewing through the experiences of a leading craft brewer.*
- Janish, Scott. The New IPA: Scientific Guide to Hop Aroma and Flavor. ScottJanish.com, 2019.
  - *Discusses the latest scientific research on hops and future trends in hop utilization.*
- Brewers Association and American Homebrewers Association websites for articles, forums, and educational resources.
  - *Provides resources and discussions on emerging trends in the brewing industry.*

Weblinks

- Brewers Association - The State of the Industry: https://www.brewersassociation.org/?s=Trends
- Craft Brewing Business - Top Trends in Craft Beer: https://www.craftbrewingbusiness.com/?s=Trends
- Beer & Brewing - Innovations in Brewing: https://beerandbrewing.com/search/?q=Future%20Trends&hPP=30&idx=dato_cbb_web&p=0&search=Future
- Brewbound - Beer Industry Trends: https://www.brewbound.com/search-results?q=Future
- Craft Beer & Brewing - Future of Brewing: https://www.craftbeer.com/?s=Future
- Beverage Daily - Beer Market Trends: https://www.beveragedaily.com/search?q=Beer&t=all&p=1&ob=date&range_date=date

These references offer valuable insights into future trends in brewing, helping brewers stay ahead and innovate in their practices.

***

Epilogue: The Brewing Odyssey

The journey of brewing weaves ancient wisdom with modern innovation, a testament to humanity's endless quest for creativity and connection, where every glass holds the past, the present, and the promise of the future.

Imagine standing in a bustling brewery, the warm, inviting aroma of malt and hops filling the air, as the soft hum of modern brewing equipment creates a symphony of progress and tradition. The journey through the art and science of brewing is a testament to humanity's enduring quest for excellence, creativity, and connection. As we conclude our exploration, it is fitting to reflect on the path we've traveled and envision the future that lies ahead.

Reflection on the Past

The story of beer began thousands of years ago, in the fertile valleys of Mesopotamia and along the banks of the Nile, where the first brewers, guided by intuition and observation, transformed humble grains into a beverage that became central to their communities. These early pioneers laid the foundation for a craft that would evolve through the ages, driven by the principles of rationalism, empiricism, and the scientific method. Their rudimentary techniques and keen observations allowed them to harness the natural fermentation processes, turning basic ingredients into a source of nourishment and pleasure. These ancient brewers, without the benefit of modern science, relied on their senses and experiences, passing down their knowledge through generations.

As we moved through time, we witnessed the monks of medieval Europe, cloistered in their abbeys, perfecting their brewing techniques and elevating beer to new heights of quality and flavor. Their meticulous records and innovative methods, such as the Bavarian Purity Law of 1516, ensured that beer remained a symbol of purity and tradition. The monks' dedication to consistency and excellence laid the groundwork for modern brewing standards. They experimented with different grains, hops, and fermentation techniques, documenting their successes and failures with precision. This era marked a significant advancement in the brewing craft, with a focus on refining the process and ensuring the highest quality of beer.

The Industrial Revolution brought technological advancements that revolutionized the brewing industry, introducing refrigeration, pasteurization, and mechanization, which enabled brewers to produce beer on an unprecedented scale while maintaining high standards of quality. The advent of steam power and automated machinery transformed breweries into large-scale operations, capable of producing consistent and reliable products for a growing consumer base. These technological innovations not only increased production capacity but also improved the safety and stability of the beer. The ability to control fermentation temperatures and preserve beer for longer periods expanded the reach of breweries and made beer a staple beverage across the globe.

Celebrating the Present

In the latter part of the 20th century, the craft beer movement emerged as a vibrant force, celebrating diversity, creativity, and the revival of traditional methods. Craft brewers, inspired by a desire to push the boundaries of flavor and innovation, embraced both ancient practices and cutting-edge technology. They experimented with new ingredients, techniques, and styles, creating a rich tapestry of beers that delighted the senses and expanded the possibilities of what beer could be. This era saw the resurgence of small, independent breweries, each with its unique approach to brewing. Craft brewers explored the vast array of flavors that different hops, malts, and yeasts could impart, leading to an explosion of new and exciting beer styles.

Throughout this journey, the principles of rationalism, empiricism, and the scientific method have been indispensable. Rationalism, with its emphasis on logical deduction and theoretical understanding, has guided brewers in developing new techniques and refining existing ones. Theoretical knowledge about yeast metabolism, fermentation kinetics, and the chemical interactions between ingredients has allowed brewers to predict outcomes and optimize their processes. Empiricism, grounded in observation and experimentation, has allowed brewers to test their ideas and learn from their experiences. By conducting controlled experiments and analyzing the results, brewers have been able to refine their methods and achieve consistent quality. The scientific method, a structured approach to inquiry, has provided a framework for systematic improvement and innovation. By observing phenomena, forming hypotheses, conducting experiments verifying or fasifying the hypotheses, analyzing data, replication, and peer review, brewers have continuously advanced their craft.

Envisioning the Future

As we look to the future, the brewing industry continues to evolve. Technological advancements such as automation and data analytics are enhancing efficiency and product quality, while emerging beer styles and consumer preferences drive innovation. Automated brewing systems allow for precise control over every aspect of the brewing process, from ingredient measurements to fermentation conditions. Data analytics provide insights into production trends and consumer preferences, enabling brewers to fine-tune their recipes and optimize their operations. Sustainability has become a central concern, with brewers seeking to reduce their environmental footprint through efficient water management, renewable energy sources, and sustainable ingredients. Initiatives such as water recycling, solar power, and the use of locally sourced, organic ingredients are becoming standard practices in many breweries. The demand for non-alcoholic and low-alcohol beers is rising, prompting brewers to develop new techniques that preserve the full flavor of traditional brews while catering to health-conscious consumers. Advanced methods such as vacuum distillation and reverse osmosis allow brewers to create flavorful, low-alcohol options without compromising on taste.

Yet, amidst all this progress, the essence of brewing remains unchanged. It is a craft that combines art and science, tradition and innovation, and individual creativity with communal celebration. It is a craft that brings people together, creating connections and fostering a sense of community. Brewing has always been about more than just the beer itself; it is about the people who make it, the places it is made, and the moments it enhances. Whether it is a small gathering of friends or a large community festival, beer has the power to bring people together and create lasting memories. It is a craft that reflects the rich history of human ingenuity and the endless possibilities of the future. Each brewer, with their unique approach and perspective, contributes to the ongoing story of beer, adding their chapter to a narrative that spans millennia.

A Timeless Craft

Imagine, once again, the bustling brewery. The brewer stands at the helm, blending ancient wisdom with modern technology, guided by the principles that have shaped the craft for millennia. The brew kettle hums, the fermenters bubble, and the aroma of hops fills the air. Each batch of beer is a testament to the brewer's skill, creativity, and dedication, a reflection of the journey that began thousands of years ago and continues to evolve. The brewer carefully monitors the process, making adjustments as needed, always striving for perfection. They draw on their knowledge of chemistry and microbiology, but also rely on their senses and intuition. The art of brewing is in the balance between precision and creativity, between science and tradition.

As we conclude this exploration of the art and science of brewing, we celebrate the rich tapestry of history, the vibrant creativity of the present, and the boundless potential of the future. Whether you are a novice homebrewer, a seasoned professional, or simply a lover of beer, may this book inspire you to embrace the journey, to explore the craft with curiosity and passion, and to raise a glass to the timeless tradition and endless innovation that define the world of brewing. The story of beer is far from over. It is a living, breathing narrative, shaped by each brewer's hands, informed by each sip's taste, and enriched by each shared moment over a pint. The art and science of brewing continue to evolve, driven by the same principles that guided our ancestors and will inspire future generations. As we look ahead, we are reminded that the journey is as important as the destination, and the best is yet to come.

In this vision of the future, the brewing industry will continue to embrace new technologies and techniques, always with an eye toward sustainability and quality. Brewers will push the boundaries of flavor, experimenting with new ingredients and methods, while staying true to the principles that have guided the craft for centuries. They will continue to build communities, bringing people together to share in the joy of beer. The future of brewing is bright, filled with endless possibilities and the promise of continued innovation and excellence.

So, as you stand in that bustling brewery, take a moment to appreciate the journey. From the ancient brewers of Mesopotamia to the monks of medieval Europe, from the industrial pioneers to the craft beer revolutionaries, each has contributed to the rich tapestry of brewing. And now, it is your turn to be a part of this story, to explore, to innovate, and to share your passion with the world. Raise your glass to the past, present, and future of brewing, and to the endless possibilities that lie ahead.

Cheers!

To brew is to shape nature with care, intention, and humility.
It is to steward the past, refine the present, and offer the future something better.
May every glass poured be a gesture of gratitude—

not only for what we create, but for how we choose to create it.

***

Glossary

From abbey brewing to zymurgy, the lexicon of beer is a testament to the craft's rich heritage and endless innovation.

A

Abbey Brewing: Brewing practices refined by monks in medieval Europe, characterized by meticulous record-keeping and innovative methods. Abbey brewing contributed significantly to the quality and diversity of beers, emphasizing tradition and purity.

Acetaldehyde: A compound that can produce a green apple or raw pumpkin flavor in beer, typically resulting from incomplete fermentation.

Acid Rest: A stage in the mashing process where the temperature is held between 95-113°F (35-45°C) to activate phytase enzymes, which break down phytin in the malt into phytic acid and minerals. This reduces the pH of the mash, improving enzyme activity, protein degradation, and overall beer clarity and stability.

Acrospire: The sprout of a germinating barley grain, whose length indicates the modification level during the malting process.

Activated Carbon Filters: Used in brewing to remove chlorine, chloramine, and impurities from water, enhancing beer taste and quality. These filters adsorb contaminants onto the activated carbon's porous surface, ensuring cleaner, consistent brewing water.

Adjuncts: Unmalted grains or other fermentable materials added to the mash to supplement malted barley, such as corn, rice, wheat, oats, and rye.

Aeration Stone: A device used to introduce oxygen into wort before fermentation, ensuring proper yeast growth and activity.

Aeration: The process of introducing oxygen into the wort before fermentation to help yeast build cell walls and reproduce. Essential during the initial yeast growth phase but should be avoided after fermentation starts to prevent oxidation.

Aging: The process of storing beer to develop and mature its flavors over time.

Alchemist: In brewing, an alchemist is a brewer who seamlessly blends scientific knowledge with creative artistry to transform basic ingredients—water, malt, hops, and yeast—into beer. This term highlights the brewer's skill in manipulating chemical processes and techniques to produce unique and flavorful beverages, much like the historical alchemists who sought to turn base materials into gold, but this time, the brewer is truly successful.

Alchemy: In brewing, alchemy refers to the artful and scientific transformation of basic ingredients—water, malt, hops, and yeast—into beer. It embodies the brewer's ability to blend precise chemical processes with creative techniques to craft unique and flavorful golden brews, successfully achieving what historical alchemists sought to do: transform base materials into gold.

Alpha Acids: Compounds in hops that contribute to bitterness in beer through isomerization during boiling. They are less affected by dry hopping, which focuses on aroma and flavor.

Alcohol: In brewing, alcohol refers to ethanol, produced by yeast during fermentation as they convert sugars into alcohol and carbon dioxide. It is measured as a percentage by volume (ABV) and influences the beer's flavor, body, and sensory experience. The fermentation process and fermentable sugar content determine the final alcohol level.

Alpha-Amylase: An enzyme active at a temperature of up to 159 F (158-162°F or 70-72°C), and a pH of 5.7, which breaks down complex starches into a mixture of fermentable sugars and dextrins, make a less fermentable wort and, consequently, a thicker, maltier beer.

American Oak: Barrels made from *Quercus alba*, commonly known as white oak used for barrel aging beer. It has more porous grain structure compared to French and Hungarial oak, allowing for faster extraction of flavors. Known for imparting bold flavors like vanilla, coconut, and sweet spices to beer. The heavy charring of American oak barrels caramelizes the wood sugars, adding a rich, smoky depth. Native to eastern and central regions of North America, e.g., Missouri, Kentucky, and Tennessee.

Amphorae: Ancient clay pots with two handles and a narrow neck used for fermenting, aging, and storing beer, imparting distinct flavors due to their porous clay material. Modern craft brewers are reviving their use for the unique characteristics they bring to the beer.

Amylases: Enzymes present in malt that break down starches into fermentable sugars during the mashing process.

Aroma Hops: Hops added late in the brewing process, including dry hopping, to impart strong aromatic qualities without significantly increasing bitterness.

Attenuation: The percentage of sugars that yeast converts to alcohol and carbon dioxide during fermentation. High attenuation results in a drier beer, while low attenuation results in a sweeter beer.

Autolysis: The self-digestion of yeast cells that occurs when they die and break down, potentially causing off-flavors such as a meaty or rubbery taste.

Automation: The use of advanced technology to control brewing processes with minimal human intervention. Automation enhances efficiency, consistency, and quality in modern brewing operations.

B

Backflushing: A cleaning technique where cleaning solution is forced through equipment in the opposite direction of normal use to dislodge and remove deposits more effectively.

Barley: A cereal grain that is the primary source of malt for brewing beer, known for its enzyme-rich endosperm.

Barrel: In brewing, a barrel is a cylindrical container, traditionally made of oak, used for aging beer. Barrels impart unique flavors and aromas, such as vanilla, caramel, and toast, through interactions between the wood and the beer. This aging process can enhance the complexity and character of the beer. Barrels are often used for styles like stouts, porters, and sour ales. They can also be previously used barrels from wine or spirits production, adding additional layers of flavor from the previous contents. Proper maintenance and sanitation of barrels are crucial to avoid contamination and spoilage.

Barrel Aging: The process of maturing beer in wooden barrels, which imparts complex flavors from the wood and any previous contents of the barrel. This technique enhances the depth, character, and aroma of the beer.

Batch Sparging: A lautering method where wort is completely drained from the mash tun, sparge water is added in a single batch, stirred, and then drained again. It is popular among homebrewers for its simplicity and ease of use.

Bavarian Purity Law—Reinheitsgebot (ryne-HYTS-guh-boht)**:** A regulation enacted in 1516 by Duke Wilhelm IV of Bavaria, mandating that beer could only be made from barley, water, and hops to ensure purity and quality.

Beta-Amylase: An enzyme most active at a temperature of 149 F (140-149°F or 60-65°C), and a pH of 5.3-5.4, which breaks down starches into maltose, a highly fermentable sugar, resulting in a more fermentable wort. Beta amylase is deactivated more rapidly at higher temperatures.

Biochemical: Biochemical refers to the chemical processes and reactions within living organisms that are essential to brewing. This includes the breakdown of starches into sugars during mashing, yeast fermentation converting sugars into alcohol, and the development of flavors and aromas.

Biodegradable: Materials that can be broken down by natural processes, typically by microorganisms, into harmless substances. Biodegradable alternatives to single-use plastics help reduce environmental impact.

Bittering Hops: Hops added at the beginning of the boil to contribute primarily to the beer's bitterness. The long boil time ensures maximum isomerization of alpha acids, which impart the desired bitterness to the beer.

Bitterness: The sharp flavor in beer primarily derived from the isomerization of alpha acids during the boil. Dry hopping typically does not contribute significantly to bitterness.

Blending: The practice of mixing barrel-aged beer with fresh beer or combining different batches of barrel-aged beer to create a balanced and harmonious final product with enhanced complexity.

Boil Duration: The length of time the wort is boiled, typically ranging from 60 to 90 minutes, affecting wort concentration, hop utilization, and flavor development.

Boil-Over: A condition during the wort boil where the foam rises and spills over the sides of the kettle, potentially causing a mess and loss of wort.

Boiling: The process of heating the wort to a rolling boil to sterilize it, extract bitterness and aroma from hops, coagulate proteins, and concentrate the wort.

Bottling: The process of transferring beer into bottles for storage and consumption. Bottling involves careful sanitization, priming, and sealing to ensure the beer remains fresh and carbonated.

Brettanomyces ("Brett"): A wild yeast known for producing unique flavors in beer, ranging from fruity and funky to earthy and barnyard-like. It can ferment complex sugars that other yeasts cannot, adding depth and complexity to sour beers.

Brewing Logs: Detailed records kept by brewers that track ingredients, temperatures, times, and deviations during the brewing process to help identify patterns and improve techniques.

Brewing Salts: In brewing, brewing salts are minerals added to water to adjust its chemical composition, enhancing the brewing process and the final beer's flavor profile. Common brewing salts include calcium sulfate (gypsum), calcium chloride, magnesium sulfate (Epsom salt), and sodium bicarbonate (baking soda). These salts help control pH levels, promote yeast health, and influence the taste and mouthfeel of the beer. Proper use of brewing salts ensures that the water chemistry matches the desired beer style, optimizing enzyme activity during mashing and overall beer quality.

Brewing Software: Digital tools used by brewers to design recipes, calculate efficiency, predict outcomes, and record brewing data.

Buffer: Buffers in brewing are chemical substances that stabilize the pH level of the mash and wort by neutralizing acids and bases. Common buffers include sodium bicarbonate (raise pH if it gets too low), and phosphates.

Burton-on-Trent: A town in England known for its high sulfate water, ideal for brewing pale ales, and influential in the development of water treatment techniques.

C

Calcium: A mineral in brewing water that helps lower wort pH, assists enzyme activity, and supports yeast health. Ideal levels are between 50-100 ppm.

Calcium Carbonate: In brewing, calcium carbonate is a mineral used to adjust the pH of the mash, increasing alkalinity when the water is too acidic. It is commonly added to the mash to raise pH levels, with a recommended maximum of 2 tsp/5 gallons of water.

Calcium Chloride: A brewing salt used to add calcium and chloride ions to water, enhancing malt sweetness and fullness. Add to mash to lower pH with a max. of 1 tsp/5 gal.

Calcium Sulfate (Gypsum): A brewing salt used to increase calcium and sulfate levels in water, enhancing hop bitterness and dryness. Add to mash to lower pH with a max. of 1 tsp/5 gal.

Campden Tablets (Potassium Metabisulfite): Tablets used to neutralize chlorine and chloramine in brewing water.

Cane Sugar (Sucrose): Common table sugar used in brewing, which can impart a cidery taste if overused but adds complex flavors when partially refined.

Carbonation: The process of dissolving carbon dioxide (CO2) into beer, creating bubbles and giving the beer its fizzy texture. Carbonation can be achieved naturally through priming or artificially through force carbonation.

Carboy: A large, narrow-necked container used in brewing for fermenting and storing liquids. Made of glass or plastic, carboys are essential for homebrewers and come in various sizes, typically from 1 to 6.5 gallons. Glass carboys are non-reactive, preserving brew purity, while plastic ones are lighter and less fragile. Often used during secondary fermentation to let beer clear and develop flavor without oxygen exposure, proper sanitation and handling are crucial to prevent contamination.

Caryophyllene: An essential oil in hops contributing to spicy and woody aromas. Its volatility makes it ideal for dry hopping to preserve these delicate scents.

Caustic Potash (Potassium Hydroxide): A strong alkaline cleaner similar to caustic soda, used in the brewing industry for cleaning equipment. It is effective at breaking down organic materials but requires careful handling due to its corrosive nature.

Caustic Soda (Sodium Hydroxide): A powerful alkaline cleaner used in the brewing industry to dissolve organic materials such as proteins and fats. It requires strict safety measures due to its corrosive nature.

Channeling: A condition during sparging where water flows through paths of least resistance in the grain bed, leaving pockets of unextracted sugars and reducing lautering efficiency.

Chemicals for Barrel Sanitization: Substances such as sulfur dioxide or potassium metabisulfite used to sanitize barrels by killing microorganisms. These chemicals must be thoroughly rinsed out before use to prevent residual flavors from affecting the beer.

Chemistry: In brewing, chemistry involves the study and application of chemical processes that convert raw ingredients into beer. This includes the bre,.akdown of starches into sugars by enzymes, the fermentation of sugars by yeast to produce alcohol and carbon dioxide, and the influence of water chemistry on flavor and clarity. Understanding these chemical interactions allows brewers to manipulate variables like pH, temperature, and ingredients to achieve desired beer characteristics.

Chill Haze: Haze that forms in beer when it is chilled, caused by proteins and polyphenols binding together.

Chloramine: A combination of chlorine and ammonia used to disinfect water, which can also cause off-flavors if not removed.

Chloride: A mineral that enhances malt sweetness and fullness in beer. Ideal levels are 50-150 ppm.

Chlorinated Trisodium Phosphate (TSP): A cleaner-sanitizer that combines the cleaning power of TSP with the biocidal properties of chlorine. It is effective but can be corrosive and is used with caution.

Chlorine: A disinfectant added to municipal water supplies that can cause off-flavors in beer if not removed.

Chlorine Dioxide: A gas used in public water treatment facilities. Similar to bleach it kills bacteria, viruses, and fungi.

Citric Acid Cycle (Krebs cycle): Also known as the Krebs cycle or TCA cycle, it is a series of enzymatic reactions in yeast and other organisms crucial for cellular respiration. In brewing, it is the metabolic pathway where yeast converts sugars into energy, producing compounds that influence beer's flavors and aromas. This cycle occurs in the mitochondria of yeast cells and is essential for efficient energy production and yeast health during fermentation.

Clean-In-Place (CIP): A method used to clean the interior surfaces of pipes, vessels, process equipment, filters, and associated fittings without disassembly. This system circulates cleaning solutions through a closed loop to ensure thorough cleaning of complex equipment.

Closed Fermentation: In brewing, closed fermentation refers to the process of fermenting beer in a sealed vessel that prevents exposure to outside air. This method helps maintain a controlled environment, protecting the beer from contamination by wild yeasts, bacteria, and oxygen, which can cause off-flavors and spoilage. Closed fermentation vessels are typically equipped with airlocks or blow-off tubes to allow carbon dioxide to escape while preventing air from entering. This process is especially crucial for brewing lagers and other styles that require precise control over fermentation conditions. It also aids in maintaining carbonation levels and ensuring a cleaner, more consistent final product.

CO2 Purging: The process of using carbon dioxide to displace oxygen in bottles or kegs to prevent oxidation and preserve beer quality.

Cold Conditioning: Cold conditioning is the process of storing beer at low temperatures (32°F to 40°F or 0°C to 4°C) after fermentation. This step improves clarity, stability, and flavor by encouraging the precipitation of unwanted proteins, tannins, and yeast. It also allows flavors to mellow and mature, resulting in a smoother taste. This process is crucial for lagers, which require longer conditioning times at colder temperatures compared to ales.

Cold Crashing: The process of lowering the temperature of beer near freezing before packaging to help clear haze by causing proteins and yeast to settle out.

Collaboration: Working jointly with other brewers on projects to share knowledge, learn new methods, and gain feedback, fostering community and innovation in brewing.

Community Engagement: Efforts to involve and educate the local community about sustainable practices, such as hosting educational events and partnering with local farms. Engaging the community fosters support and awareness of sustainability initiatives.

Composting: The process of recycling organic waste, like spent grain, into nutrient-rich soil. Composting spent grain reduces waste and supports local agriculture by providing valuable resources for farming.

Contact Time: The duration hops remain in contact with the beer during dry hopping. Longer contact times can intensify aromas but may introduce grassy or vegetal flavors.

Continuous Improvement: The ongoing practice of refining brewing techniques and processes to achieve consistently high-quality results, often through meticulous record-keeping, experimentation, and learning.

Coolship: A shallow, open fermentation vessel traditionally used in brewing, especially for spontaneously fermented beers like Lambics. Its wide surface area allows hot wort to cool quickly while being exposed to wild yeasts and bacteria from the environment, initiating fermentation.

Coriander: A spice used in brewing, particularly in styles like Gose, to add citrusy and spicy notes.

Corn: In brewing, corn is used as an adjunct to add fermentable sugars, lighten the body, and reduce costs. It provides a mild sweetness and enhances clarity and crispness, commonly used in American lagers and light beer styles. Corn is added as grits, syrup, or flaked maize. Proper incorporation avoids off-flavors and achieves desired results.

Corn Sugar (Dextrose): A highly fermentable sugar often used for priming beer, adding alcoholic strength without residual sweetness.

Counterflow Wort Cooler: A piece of equipment used to rapidly cool wort. These coolers have hidden surfaces that require meticulous cleaning with powerful chemicals to prevent microbial contamination.

Crabtree Effect: A phenomenon in yeast fermentation where, in the presence of high glucose concentrations, yeast cells ferment sugar rather than respire, even when oxygen is available. This leads to immediate alcohol production with low energy release, disrupting normal cell growth and energy storage. As a result, high-sugar beers can experience weak and prolonged fermentations.

Craft Beer Movement: A resurgence of small-scale brewing in the late 20th century that emphasizes diversity, innovation, and traditional methods, often in contrast to large commercial breweries.

Crown Cap: A metal cap used to seal beer bottles, preventing contamination and maintaining carbonation. Crown caps are crimped onto the bottle using a capping tool.

Cryo Hops: A concentrated form of hops produced by separating lupulin glands from hop cones at very low temperatures. They provide intense hop flavor and aroma with less plant material.

D

Darcy's Law: A principle describing the flow of fluid through a porous medium, applied in brewing to design efficient lauter tuns and grain beds for optimal wort flow.

Data Analytics: The process of examining brewing data to identify patterns and correlations that can optimize brewing operations and improve product quality.

Dechlorination: The process of removing chlorine and chloramine from water to prevent off-flavors in beer.

Decoction Mashing: A traditional mashing technique where a portion of the mash is removed, boiled, and then returned to the main mash to enhance malt character, improve extraction, and increase mash temperature to the set-point.

Deductive Reasoning: A top-down logical process that starts with a general theory or hypothesis and then tests it through specific observations and experiments.

Diacetyl Rest: A period at the end of primary fermentation, typically lasting 24-48 hours, where the temperature is raised to encourage yeast to reabsorb diacetyl and other unwanted byproducts, reducing buttery off-flavors.

Diacetyl: A buttery-flavored compound produced by yeast during fermentation. A diacetyl rest allows the yeast to reabsorb and reduce this compound before fermentation is complete.

Diastatic Power: A measure of a malt's enzyme content, indicating its ability to convert starches into fermentable sugars, measured in degrees Lintner.

Dry Hopping: The process of adding hops to beer after the boil and primary fermentation to enhance aroma and flavor without adding significant bitterness.

E

Ellagitannins: Compounds found in French oak that add gentle astringency and complexity to beer. They contribute to a refined, wine-like character.

Emmer Wheat: Emmer wheat, also known as farro, is an ancient grain used in brewing to add unique flavors and historical authenticity to beer. This hulled wheat variety (Triticum dicoccum) was one of the first domesticated crops in the Near East and has a rich, nutty flavor profile. In brewing, emmer wheat can enhance the complexity of the malt profile, contributing earthy, slightly sweet, and nutty notes. It is often used in specialty beers, historical brews, and traditional styles to provide distinctive characteristics and connect modern brewing with ancient agricultural practices. Proper milling and mashing techniques are essential to fully extract its flavors and fermentable sugars.

Empiricism: Empiricism is a philosophical approach emphasizing bottom-up inductive reasoning. It relies on specific sense data and experiences to generate broad theoretical principles. Empiricists gather data through observations and experiments, using these instances to formulate general concepts. This method contrasts with rationalism, which uses top-down deductive reasoning based on definitional truths and theoretical frameworks.

Endosperm: In brewing, the endosperm is the starchy interior of a grain, such as barley. During malting, enzymes break down the endosperm's starches into fermentable sugars. In mashing, these sugars are further extracted and converted, providing the essential nutrients for yeast to ferment and produce alcohol and carbonation in beer.

Energy-Efficient Brewing Systems: Brewing equipment and technologies designed to minimize energy consumption, such as heat recovery systems and energy-efficient boilers. These systems help reduce a brewery's carbon footprint and operational costs.

Enzymes: Proteins that catalyze biochemical reactions, such as the breakdown of starches into sugars during mashing.

Enzyme Activity: The catalytic action of enzymes like alpha-amylase and beta-amylase in breaking down starches into fermentable sugars during mashing.

Epsom Salts (Magnesium Sulfate): A brewing salt used to add magnesium and sulfate to water, influencing beer flavor and mouthfeel.

Essential Oils: Volatile compounds in hops, such as myrcene, humulene, and caryophyllene, that contribute to aroma and flavor. They are best preserved through dry hopping.

Esters: In brewing, esters are aromatic compounds produced during fermentation from the reaction between organic acids and alcohols. They contribute fruity and floral notes, such as banana, apple, and pear. The formation of esters is influenced by yeast strain, fermentation temperature, and wort composition, allowing brewers to manage the ester profile of their beer.

Ethics: Moral principles that govern the conduct of individuals and organizations in the brewing industry, encompassing issues like sustainability and fair labor practices.

European Brewing Convention (EBC) Units: In brewing, EBC units measure the color of beer and malt, ranging from pale yellow to dark brown. Developed by the European Brewing Convention, the EBC value is determined by measuring light absorption at 430 nm. Higher EBC values indicate darker beers, helping brewers maintain consistency and accurately communicate beer color profiles.

Evaporation Rate: The rate at which water evaporates from the wort during boiling, affecting wort concentration and final gravity. Consistent evaporation is essential for achieving desired beer characteristics.

Experimentation: The process of trying out new techniques, ingredients, or methods in brewing to discover if the results verify or falsify a predictive hypothesis.

Extraction Efficiency: The effectiveness of transferring desired hop compounds, such as essential oils, from hops into the beer. Dry hopping focuses on maximizing aromatic oil extraction.

F

False Bottom: A perforated plate or screen at the bottom of the lauter tun that allows wort to pass through while retaining the grain husks, facilitating effective separation.

Fermentable Sugars: Sugars that can be converted by yeast into alcohol and carbon dioxide during fermentation.

Fermentation: The process by which yeast converts fermentable sugars in wort into alcohol and carbon dioxide, producing beer.

Fermentation Kinetics: In brewing, fermentation kinetics is the study of the rates at which yeast converts sugars into alcohol and carbon dioxide. It examines factors like temperature, yeast strain, and nutrient availability, as well as changes in substrate and product concentrations over time. Understanding fermentation kinetics helps brewers optimize conditions, improve consistency, and achieve desired beer characteristics.

Fining Agents: Substances like Irish moss, Whirlfloc tablets, or gelatin added to beer during the brewing process to help precipitate and remove haze-causing particles.

Finings: Finings are substances added to beer or wort to clarify and stabilize the final product. They bind to unwanted particles like proteins, yeast, and polyphenols, causing them to precipitate and settle. Common finings include isinglass, gelatin, carrageenan (Irish moss), and polyclar. These agents improve clarity, taste, and stability, reducing haze and extending shelf life.

Flanders Red Ale: A Belgian-style sour beer known for its deep red color and complex sour flavors, often with fruity and malty notes. It is typically aged in oak barrels, which contribute to its depth and complexity through blending younger and older beers.

Flavor Development: The process of enhancing the taste and aroma of beer through various brewing techniques, ingredients, and aging.

Flavor Hops: Hops added midway through the boil to contribute both bitterness and flavor. The shorter boil time preserves more of the hop's essential oils, adding complex flavors to the beer.

Flocculation: The clumping together and settling out of yeast cells after fermentation, affecting the clarity and yeast character of the beer.

Floral: A common aroma descriptor for hops, indicating scents reminiscent of flowers. Many hop varieties, including Cascade and Saaz, are known for their floral characteristics.

Fly Sparging: A continuous sparging method where sparge water is added to the mash tun while wort is simultaneously drained from the bottom, maximizing sugar extraction.

Force Carbonation: A method of carbonating beer by directly infusing CO2 into a keg under pressure, allowing for precise control over carbonation levels and quicker results compared to natural carbonation.

French Oak: Barrels made from *Quercus robur*, known as European or English oak, used for barrel aging beer. Known for its fine grain and tighter pores compared to American oak, which results in slower transfer of flavors into the beer. Known for imparting subtle and refined flavors such as gentle tannins, soft vanilla, and nuanced spice notes to beer. French oak is often used for aging lighter, more delicate beer styles. Native to Europe, e.g., France, Germany and the UK.

Furfural: A compound found in Hungarian oak that adds rich, toasty notes to beer, enhancing its complexity and depth.

Fusel Alcohols: In brewing, fusel alcohols are higher alcohols produced during fermentation, especially at higher temperatures. They contribute strong, solvent-like flavors and aromas, described as hot, spicy, or alcoholic. While small amounts add complexity to some beer styles, excessive fusel alcohols are undesirable. Controlling fermentation temperature and yeast health is crucial to managing fusel alcohol production.

G

Gelatin: In brewing, gelatin is a fining agent used to clarify beer by removing suspended particles. Derived from animal collagen, it is typically dissolved in warm water and added to the cold beer. Gelatin works by binding to proteins and other haze-forming compounds, causing them to settle to the bottom. This process improves the clarity and stability of the final product, resulting in a clearer beer with reduced haze.

Gelatinization: The process of heating starches in adjuncts like rice or corn to make them soluble and accessible to enzymatic breakdown during mashing.

Germination: In brewing, germination is the stage during malting where barley grains are soaked in water to sprout. This process activates enzymes that convert starches into fermentable sugars, essential for brewing. Germination is halted by drying and kilning the malt, preserving the enzymatic activity needed for mashing.

Glycolysis: A metabolic pathway that converts glucose into pyruvate, yielding ATP and NADH, and leading to ethanol and carbon dioxide production during fermentation.

Gose: A German-style sour beer brewed with coriander and salt, known for its refreshing tartness and slight saline note. It undergoes lactic fermentation, which imparts a tangy flavor.

Grain Bed: The layer of crushed malt and adjuncts in the lauter tun during lautering, which acts as a natural filter to separate wort from solid grain husks.

Grain Sugar: Grain sugar refers to sugars derived from the starches in grains like barley, corn, wheat, and rice. These sugars are produced through enzymatic conversion or acid hydrolysis and are used in brewing to provide fermentable sugars for yeast, increasing the alcohol content without significantly altering the flavor profile of the beer.

Gravity: A measure of the density of wort or beer relative to water. Original gravity (OG) measures before fermentation, and final gravity (FG) measures after fermentation to calculate alcohol content.

Green Pads: Scrubbing tools made of abrasive material, ideal for removing deposits from metal surfaces in the brewing process. They are effective but must be used carefully to avoid scratching equipment.

Grist: Crushed malt that is mixed with hot water during the mashing process to create a mash.

Gypsum: See Calcium Sulfate.

H

Hard Water: Water with high mineral content, particularly calcium and magnesium, which can affect beer flavor and brewing processes.

Heat Recovery Systems: Technologies that capture and reuse energy that would otherwise be wasted, such as capturing heat from brewing processes to preheat water.

Heat Transfer: The process of transferring heat to the wort during boiling, affecting evaporation, isomerization of alpha acids, and protein coagulation.

Hop Additions: The practice of adding hops to the boiling wort at different times to achieve desired levels of bitterness, flavor, and aroma.

Hop Bags/Cages: Devices used to contain hops during dry hopping, facilitating easier removal and minimizing the risk of oxygen ingress and contamination.

Hop Burn: A harsh, astringent sensation often associated with over-dry hopping. It can be mitigated by adjusting hop quantities and contact time.

Hop Creep: Unintended continued fermentation caused by enzymes in hops during dry hopping, leading to changes in carbonation and potential off-flavors.

Hop Extracts: Concentrated hop oils and resins used for precise control over bitterness and aroma. They offer consistency but may lack the complexity of whole or pellet hops.

Hop Oils: Volatile compounds in hops responsible for aroma and flavor. They are sensitive to heat and are best preserved through dry hopping.

Hop Schedules: The timing and quantity of hop additions during brewing, influencing bitterness, aroma, and flavor.

Hop Storage: Proper storage practices to maintain hop freshness and potency, including vacuum sealing and refrigeration or freezing to protect from oxygen and light.

Hop Utilization: The efficiency with which alpha acids in hops are isomerized during boiling, imparting bitterness to the beer. Hop utilization is influenced by boil duration, temperature, and wort composition.

Hops: Cone-shaped flowers of the Humulus lupulus plant used in brewing to add bitterness, flavor, and aroma. Hops balance malt sweetness, provide antimicrobial properties, and impart flavors ranging from floral and citrus to piney and earthy. Different hop varieties achieve specific characteristics in beer styles like Pale Ales, IPAs, and Lagers.

Hot Break (Protein Coagulation): During the boiling phase of brewing, heat causes water molecules to detach from larger proteins in the wort, a process known as denaturing. This makes the large protein molecules insoluble, leading them to flocculate (clump together) if they collide with each other. Turbid (vigorous) boiling aids in the flocculation of proteins, tannins (polyphenols), and lipids, resulting in what is known as the "hot break." This separation is crucial for the clarity and stability of the final beer. However, achieving a hot break is challenging if the wort's pH is lower than 5.0.

Humulene: An essential oil in hops contributing to earthy and woody aromas. Its stability makes it suitable for both boiling and dry hopping.

Hungarian Oak: Barrels made from *Quercus frainetto*, known as Hungarian oak, or Quercus petraea, known as sessil oak, used for barrel aging beer. Pore size and tightness of grain is somewhere between American oak and French oak, but closer to French oak. Provides a balanced combination of vanilla, spice, and tannins, suitable for a wide range of beer styles. It offers a unique flavor profile that falls closer to French oak in terms of intensity. Native to Eastern Europe, e.g., Hungary, Romania, and the Balkins.

Hybrid Beers: Beers that combine characteristics of two different styles, such as IPA-lager hybrids, which blend the hoppy flavors of IPAs with the crisp profile of lagers. These hybrids offer novel flavor experiences.

Hydrometer: An instrument used to measure the specific gravity of beer, indicating the amount of fermentable sugars and helping to determine the beer's alcohol content.

Hypochlorite (Bleach): A traditional chlorine-based sanitizer effective at killing microbes but prone to causing off-flavors and corrosion, making it less favored in modern commercial brewing.

Hypothesis Formation: The process of developing theories about potential improvements in brewing based on observations and existing knowledge.

I

Inductive Reasoning: A bottom-up logical process where generalizations are made based on observations and experimental data.

Industrial Revolution: A period of significant technological advancement in the 18th and 19th centuries that transformed brewing with innovations like the steam engine, refrigeration, pasteurization, and mechanization enabling large-scale beer production.

Iodine Test: Simple and quick procedure to check for the presence of residual starch in the mash. If the solution turns blue or black, it indicates that starches are still present and have not been fully converted to sugars.

Iodophors: Sanitizers based on iodine, effective at lower concentrations compared to chlorine-based sanitizers. They are safe for stainless steel and glass, offering a visual indicator of effectiveness by changing color when their antimicrobial activity is exhausted.

IPA-Lager Hybrids: A type of hybrid beer that combines the bold hop flavors of an IPA with the clean, crisp profile of a lager. This style appeals to consumers seeking innovative and refreshing beer options.

Irish Moss: Irish moss is a type of red seaweed (Chondrus crispus) used as a fining agent to clarify wort. Added during the last 15 minutes of the boil, it helps coagulate proteins and other suspended solids, causing them to settle and resulting in clearer beer. Irish moss improves the clarity and quality of the final product by reducing haze.

Isomerization: The chemical process by which alpha acids in hops are converted to iso-alpha acids during boiling, imparting bitterness to the beer.

K

Keg: A cylindrical container used for storing and dispensing beer. Kegs are typically made of metal and can be reused multiple times, offering a durable and efficient packaging solution.

Kegerator: A refrigerated unit designed to hold kegs and dispense beer, equipped with a tap and CO2 system to maintain serving pressure and temperature.

Kegging: The process of packaging beer in kegs, often involving force carbonation and serving through draft systems.

Kettle Souring: A method of souring the wort with Lactobacillus before boiling, allowing for faster production and greater control over the level of sourness compared to traditional methods.

Kilning: The drying process in malting that halts germination and develops the malt's flavor and color. At higher kiln temperatures, the malt's diastatic strength decreases while its color darkens and caramelization flavors intensify. Conversely, at lower kiln temperatures, the malt retains greater diastatic strength, resulting in a lighter color and milder caramelization flavors.

Krausen: The foamy head of yeast and proteins that forms on top of the wort during vigorous fermentation, indicating active yeast activity.

Krebs Cycle (Citric Acid Cycle): Also known as the citric acid cycle or TCA cycle, it is a series of enzymatic reactions in yeast and other organisms crucial for cellular respiration. In brewing, it is the metabolic pathway where yeast converts sugars into energy, producing compounds that influence beer's flavors and aromas. This cycle occurs in the mitochondria of yeast cells and is essential for efficient energy production and yeast health during fermentation.

L

Lactic Fermentation: Process wherein lactic acid bacteria convert sugars, primarily glucose, into lactic acid. This process is essential in certain beer styles like Berliner Weisse and Gose, contributing to their distinct sour flavors and enhanced preservation.

Lactobacillus: A type of bacteria that produces lactic acid, contributing a clean, tangy sourness to beer. It is commonly used in both traditional and modern sour beer production.

Lactones: Compounds in oak that contribute to coconut and woody aromas in barrel-aged beer. American oak, in particular, is high in lactones, making it a popular choice for pronounced woody character.

Lag Phase: The initial stage of fermentation where yeast acclimates to the environment, absorbs oxygen and nutrients, and prepares for cell division.

Lambic: A traditional Belgian beer brewed using spontaneous fermentation, where the wort is exposed to natural yeast and bacteria in the open air. Aged hops are used for preservation, and the beer is aged in oak barrels for extended periods.

Lauter Tun: A vessel specifically designed for separating the liquid wort from the solid grain husks during lautering, equipped with a false bottom or slotted plate.

Lautering: The process of separating the liquid wort from the solid grain husks after mashing, critical for ensuring beer clarity, flavor, and quality.

Locally Sourced Ingredients: Ingredients obtained from local farms and suppliers, reducing transportation-related emissions and supporting the local economy. Using locally sourced ingredients promotes sustainability and ensures fresher products.

Louis Pasteur: A 19th-century scientist whose research on yeast and fermentation significantly advanced brewing science, leading to better fermentation control and the development of pasteurization.

Lovibond (°L): A scale used to measure the color of malt and beer, where higher values indicate darker colors.

Low-Alcohol Beers: Beers with a lower alcohol content, typically below 3.5% ABV, that still maintain a full flavor profile. These beers cater to health-conscious consumers looking for lighter drinking options.

Lupulin: The yellowish powdery substance in hop cones containing resins and essential oils responsible for bitterness, flavor, and aroma in beer.

M

Magnesium: A mineral essential for yeast health but can impart a bitter flavor at high levels. Optimal range is 10-30 ppm.

Magnesium Sulfate: (See Epson Salt)

Maillard Reactions: Maillard Reactions: In brewing, Maillard reactions are chemical processes that occur between amino acids and reducing sugars during the malting and roasting of barley. These reactions contribute to the development of complex flavors, aromas, and colors in malt and, consequently, in beer. The Maillard reactions are responsible for creating toasty, nutty, and caramel-like flavors, as well as the rich browning of the malt. These reactions play a crucial role in defining the character and depth of many beer styles, especially those with darker malts like stouts and porters.

Malt: Barley or other grains that have been germinated and kilned, forming the base ingredient in beer brewing.

Mashing: The process of mixing crushed malt with hot water to activate enzymes that convert complex starches into fermentable sugars.

Mashing-In: The initial step of mashing, where crushed malt is combined with hot water to create a mash. Mash water should be at the precise temperature and a pH of 7 but no lower.

Mash-Out: The process of raising the temperature of the mash to around 168-170°F (75-77°C) at the end of the mashing period. This step halts enzyme activity, stabilizes the sugar profile, and makes the mash less viscous, facilitating easier lautering and extraction of wort.

Mash pH: The acidity or alkalinity of the mash, typically maintained between 5.2 and 5.6 to ensure optimal enzyme activity and flavor stability.

Mash Stiffness: Refers to the ratio of water to grain in the mash, affecting its thickness or viscosity. A stiffer mash (around 1.33 quarts of water per pound of grain) favors initial starch and protein breakdown, improving wort clarity but may inhibit further enzyme activity due to high sugar concentration. A thinner mash (approximately 2.5 quarts per pound) allows more complete carbohydrate breakdown but takes longer and can cause faster enzyme deactivation.

Mash Tun: A vessel used for mixing crushed malt with hot water during the mashing process, equipped to maintain temperature and sometimes facilitate heating and where the conversion of starches to sugars occurs.

Mechanical Stirring Devices: Equipment used to maintain even temperature distribution and prevent the grain bed from compacting during lautering.

Membrane Bioreactor: An advanced water recycling system that treats and reuses water within the brewery. Membrane bioreactors help reduce overall water consumption by recycling process water for non-potable purposes.

Mesopotamia: The ancient region where the earliest known brewing practices began around 5,000 BCE, particularly among the Sumerians.

Microbiology: The branch of science dealing with microorganisms, including yeast, critical for understanding and controlling fermentation in brewing.

Mineral Content: The levels of various minerals in brewing water, which influence the flavor, mouthfeel, and stability of the beer.

Mixed Fermentation: The use of multiple yeast strains and/or bacteria to ferment beer, often resulting in complex and unique flavor profiles.

Modular Systems: Brewing systems that can be expanded or reconfigured as needed, providing flexibility and scalability for breweries. Modular designs support growth and adaptation to changing market conditions.

Monastic Breweries: Breweries operated by monks in medieval Europe, known for their systematic and innovative approaches to brewing, contributing to the advancement of brewing techniques.

Myrcene: An essential oil in hops contributing to citrusy, piney, and herbal aromas. Its high volatility makes it ideal for dry hopping to capture these delicate scents.

N

Natural Carbonation: The process of carbonating beer by adding priming sugar before sealing it in bottles or kegs. The sugar ferments, producing CO2 and carbonating the beer naturally over time.

Networking: Building relationships with other brewers and industry professionals to gain insights, share knowledge, and collaborate on projects.

Ninkasi: The Sumerian goddess of beer, celebrated in the "*Hymn to Ninkasi*," which includes an ancient brewing recipe and highlights the cultural significance of beer in Mesopotamia.

Nitrogen Content: An essential element in brewing found in proteins, which is used to estimate total protein content in malt (total protein content = 6.25 times total nitrogen content). Higher modification of malt increases soluble nitrogen, improving protein modification and reducing haze in beer. Lower total nitrogen content generally leads to clearer beer with less risk of haze formation.

Non-Alcoholic Beers: Beers that contain little to no alcohol, typically below 0.5% ABV, made using techniques like vacuum distillation or reverse osmosis to remove alcohol while preserving flavor.

O

Oats: An adjunct used to add body and creaminess to beer, commonly found in stouts.

Observation: The act of monitoring brewing processes and outcomes to identify areas for improvement and gather data for analysis.

Off-Flavor: Unwanted flavors in beer caused by various factors, including contamination, fermentation issues, or improper ingredient handling.

Organic Brewing: Brewing that uses ingredients grown without synthetic pesticides, herbicides, and fertilizers. Organic brewing practices reduce chemical runoff and promote soil health, offering environmental and health benefits.

Over-Pitching: Adding too much yeast to the wort, which can suppress the development of complex esters and phenols, leading to a bland beer.

Oxidation: A chemical reaction involving oxygen that can lead to off-flavors in beer, such as stale or cardboard-like tastes.

Ozone Treatment: A modern method of barrel sanitization that uses ozone gas to kill bacteria, yeast, and mold without leaving any residue, ensuring the barrel is free of contaminants.

P

Packaging: In brewing, packaging is the final step where beer is transferred into bottles, cans, or kegs for storage and distribution. This process ensures beer quality, prevents contamination, and maintains carbonation. Key elements include sanitation, filling, sealing, and labeling, which preserve flavor, aroma, and shelf life while protecting the beer from light and temperature changes.

Pale Ale Malt: A type of malt with a rich, biscuity flavor, commonly used in traditional British ales.

Pasteurization: A process developed by Louis Pasteur to heat liquids to kill unwanted microorganisms, extending the shelf life and safety of beer without affecting flavor.

Pediococcus: A type of bacteria that produces intense sourness and can develop diacetyl, adding a buttery flavor. It is used in sour beer production, requiring careful management to balance its characteristics.

Peer Review: The assessment of brewing techniques, experiments, and results by other brewers or experts to validate findings and improve practices.

Pellet Hops: Hops processed into small pellets for ease of storage and use. Pellet hops offer better stability and higher utilization rates compared to whole hops.

Peroxyacetic Acid: A modern, oxygen-based sanitizer known for its non-corrosive properties and suitability for no-rinse applications. It is gaining favor in commercial breweries for its effectiveness and safety.

pH (Potential of Hydrogen): In brewing, pH is a measure of the acidity or alkalinity of a solution, on a scale from 0 to 14. It is critical in various stages of brewing, including mashing, fermentation, and finishing. Proper pH levels ensure optimal enzyme activity during mashing, promote yeast health and efficient fermentation, and enhance the stability and flavor of the final beer. Monitoring and adjusting pH is essential for producing high-quality beer. Optimal range for brewing water is 5.2-5.6.

Phenolic: A type of off-flavor that can taste medicinal, smoky, or spicy, often caused by wild yeast or bacteria contamination or high fermentation temperatures.

Phenols: In brewing, phenols are chemical compounds produced during fermentation that impart flavors and aromas like spicy, smoky, clove, medicinal, and band-aiksd. Influenced by yeast strain, fermentation conditions, and raw materials, phenols are desirable in some beer styles (e.g., Belgian ales) but can be off-flavors in others. Proper brewing management helps control phenol levels.

Piney: A hop-derived aroma or flavor reminiscent of pine needles or resin. Common in hop varieties like Chinook, Simcoe, and Columbus, this characteristic adds a sharp, refreshing quality to beers, especially in American Pale Ales and IPAs, enhancing their complexity and depth.

Pitching: The act of adding yeast to the wort to start fermentation. Proper pitching rates are essential for healthy and efficient fermentation.

Polyphenols: In brewing, polyphenols are compounds in malt and hops that influence beer's flavor, stability, and clarity. They contribute to astringency and bitterness and have antioxidant properties. Polyphenols can form complexes with proteins, causing haze. Managing their levels ensures a balanced and stable beer.

Potassium: A mineral that can inhibit certain enzymes in the mash and should be used cautiously in brewing water.

Potassium Metabisulfite: See Campden Tablets

Practical Applications: The real-world implementation of brewing techniques and methods based on theoretical knowledge and empirical evidence to achieve desired results.

Primary Fermentation: The initial stage of fermentation where the majority of sugar conversion occurs, producing alcohol and carbon dioxide.

Priming Sugar: Sugar added to beer before bottling to induce fermentation and natural carbonation. Common types include corn sugar (dextrose), cane sugar (sucrose), and dry malt extract (DME).

Protein Coagulation (Hot Break): During the boiling phase of brewing, heat causes water molecules to detach from larger proteins in the wort, a process known as denaturing. This makes the large protein molecules insoluble, leading them to flocculate (clump together) if they collide with each other. Turbid (vigorous) boiling aids in the flocculation of proteins, tannins (polyphenols), and lipids, resulting in what is known as the "hot break." This separation is crucial for the clarity and stability of the final beer. However, achieving a hot break is challenging if the wort's pH is lower than 5.0.

Protein Rest: A step/rest in step mashing at a temperature around 122°F (50°C) where proteolytic enzymes break down proteins into amino acids and peptides. An over-long protein rest can cause problems with foam retention and a thin body, so if you choose to include a protein rest, keep it to half an hour or less. The pH of a typical mash (5.3) is very close to optimum conditions for proteolysis when temp pH is considered. (Note: At room temp, 5.5

will translate into 5.3 at 122 F because pH values drop slightly with an increase in temperature.)

Purging: The process of removing oxygen from bottles or kegs by filling them with CO2 before sealing. Purging helps preserve beer freshness and prevent oxidation.

R

Rainwater Harvesting: Collecting and using rainwater for non-potable purposes in the brewing process to reduce freshwater consumption.

Rationalism: A philosophical approach that uses top-down deductive reasoning. It relies on definitional truths and broad theoretical principles, applying these general concepts to specific instances. This method ensures practices are grounded in logical and theoretical frameworks, contrasting with empiricism, which focuses on observation and experimentation to build knowledge.

Recipe Development: The process of designing and refining beer recipes by experimenting with different ingredients and techniques to achieve unique flavor profiles.

Record-Keeping: The detailed documentation of every step in the brewing process to help track and analyze variables, identify patterns, and improve techniques.

Recycling: The process of converting waste materials into new materials and objects. Recycling packaging materials like glass, aluminum, and cardboard reduces waste and conserves resources in brewing operations.

Refractometer: An instrument used to measure the specific gravity of wort or beer by determining the refraction of light, useful for monitoring fermentation progress.

Refractory: Resistant to decomposition or marked by unresponsiveness; in brewing, typically refers to materials or processes that are difficult to alter or influence.

Reinheitsgebot (ryne-HYTS-guh-boht) **Bavarian Purity Law:** A regulation enacted in 1516 by Duke Wilhelm IV of Bavaria, mandating that beer could only be made from barley, water, and hops to ensure purity and quality.

Renewable Energy: Energy from sources that are naturally replenishing, such as solar and wind power. Using renewable energy in brewing operations helps reduce greenhouse gas emissions and dependence on fossil fuels.

Repeatability: The ability to obtain consistent results using the same methods, procedures, and conditions within the same brewery or laboratory.

Replicability: The ability of others to obtain consistent results using the same methods and procedures in different breweries or laboratories, by different brewers.

Reverse Osmosis (RO): A process used in brewing to remove alcohol from beer, resulting in non-alcoholic versions that retain the original flavor. Reverse osmosis is a preferred method for creating high-quality non-alcoholic beers.

Rice Hulls: An adjunct added to the mash to increase porosity and prevent stuck sparges by creating a more porous grain bed.

Rice: A commonly used adjunct in brewing that imparts a light, clean flavor, often used in mass-produced lagers.

Rule of St. Benedict: The Rule of St. Benedict, from the 6th century, emphasizes a balanced life of prayer, work, and community. Monks brewed beer to support their communities and welcome guests, adhering to principles of diligence and hospitality. This tradition of quality brewing continues in many monastic breweries today.

S

Saccharification: The enzymatic conversion of complex starches into simpler fermentable sugars during the mashing process.

Salts: In brewing, salts refer to various mineral compounds that can be dissolved in water to adjust its chemical profile. Common brewing salts include calcium sulfate (gypsum - lower pH and enhance hop bitterness), calcium chloride (lower pH and enhance sweetness), magnesium sulfate (Epsom salts - yeast nutrient).

Sanitization: The practice of cleaning and disinfecting brewing equipment to eliminate unwanted microorganisms and prevent contamination that may cause beer off-flavors.

Scalability: The ability of a brewing system or process to be expanded to accommodate increased production demands without compromising quality.

Scientific Method: A systematic approach to research involving observation, hypothesis formation, experimentation for verfying or falsifying the hypothesis, data analysis with assessment, replicability, and peer review. The scientific method drives innovation and refinement in brewing techniques.

Secondary Fermentation: A slower stage of fermentation after primary fermentation, where remaining complex sugars are converted, and the beer matures and clarifies.

Sensory Analysis: The practice of evaluating beer characteristics, such as aroma, flavor, and mouthfeel, through sensory methods to refine brewing techniques.

Serving Pressure: The pressure at which beer is dispensed from a keg, typically maintained between 8-12 psi. Serving pressure affects carbonation levels and the quality of the pour.

Single Infusion Mash: A mashing technique where the mash is held at a constant temperature throughout the process, suitable for well-modified malts.

Single-Use Plastics: Plastics designed to be used once and then discarded, such as straws and cups. Reducing single-use plastics in brewing operations decreases environmental pollution and waste.

Six-Row Brewers' Malt: A malt type with higher enzyme content and protein levels, ideal for converting adjuncts like corn and rice. It has a thicker husk, aiding in efficient sparging but can increase tannin extraction. Commonly used in American lagers and light beers, it enhances fermentability and adds a grainy flavor.

Soap: Soaps are cleaning agents composed of natural fats and oils combined with alkaline substances. They contain surfactants, which reduce water's surface tension, enabling the emulsification and removal of organic residues like proteins and fats. Soaps are used to clean brewing equipment, but they must be thoroughly rinsed off to prevent any residue from affecting beer flavor and head retention. While effective, soaps are often less efficient than specialized brewing cleaners and sanitizers, which are designed to meet the specific cleaning needs of brewing environments.

Sodium: A mineral that adds smoothness and enhances mouthfeel in beer. Optimal levels are 50-150 ppm.

Soft Water: Water with low mineral content, often preferred for brewing certain beer styles like lagers.

Solar Panels: Devices that convert sunlight into electricity, used to power brewing operations or heat water in the brewing process. Solar panels provide a renewable energy source that reduces a brewery's carbon footprint.

Sour Beers: Beers characterized by a tart and acidic flavor, often produced through mixed fermentation with wild yeast and bacteria.

Sour Stouts: A hybrid beer style that combines the rich, roasted flavors of stouts with the tartness of sour beers.

Sparge Water: Hot water used to rinse the grain bed during lautering to extract remaining sugars from the mash. Acidify the sparge water to a pH of 5.7

Sparging: The process of rinsing the grain bed with hot water during lautering to extract the remaining sugars from the mash. Acidify the sparge water to a pH of 5.7

Specific Gravity: A measure of the density of beer relative to water, used to determine the amount of fermentable sugars and the beer's alcohol content. Specific gravity is measured with a hydrometer or refractometer.

Spent Grain: The leftover grain from the brewing process, which can be composted or used as animal feed.

Spontaneous Fermentation: A fermentation process where the wort is exposed to the open air, allowing natural yeast and bacteria to inoculate it. This method is traditionally used in the production of Lambics.

Steaming: A method of sanitizing barrels by using high-pressure steam to kill bacteria and other microorganisms, ensuring the barrel is clean without introducing chemicals.

Steeping: In malting, steeping is the initial process where raw grains, typically barley, are soaked in water to initiate germination. This process hydrates the grains, activates enzymes, and prepares them for the subsequent germination and kilning stages.

Step Mashing: A mashing technique involving gradually increasing the mash temperature through a series of steps/rests to activate different enzymes at their optimal temperatures.

Sterilization: Refers to the process of eliminating all microorganisms from equipment, containers, and surfaces to prevent contamination. This is achieved through methods such as heat, chemical agents, or steam. Proper sterilization ensures that only the desired yeast strains ferment the wort, resulting in a clean and uncontaminated beer.

Stoneware Bottle: An early type of beer container made from stoneware, used before the widespread adoption of glass bottles. Stoneware bottles offered limited control over carbonation and preservation.

Strike Temperature: The temperature of the water added to the crushed malt during mashing-in, usually higher than the desired mash temperature to account for heat loss from the cooler malt.

Stuck Fermentation: A situation where fermentation stops prematurely, often due to poor yeast health, insufficient nutrients, or temperature fluctuations.

Stuck Mash: A situation where the mash becomes compacted, hindering the flow of wort during lautering and sparging, often due to improper grain crush or bed depth.

Stuck Sparge: A condition where the grain bed compacts, preventing wort from flowing through, often caused by improper grain crush or bed depth.

Sucrose: See Cane Sugar.

Sulfate: In brewing, sulfate (SO_4^{2-}) is a mineral ion that enhances the perceived dryness and bitterness of beer, making it beneficial for hoppy styles like IPAs, where levels above 200 ppm are common. It is added through salts such as calcium sulfate (gypsum) and magnesium sulfate (Epsom salt). Proper sulfate levels help achieve the desired flavor balance and hop character, while excessive amounts can cause harshness.

Sumerians: An ancient civilization in Mesopotamia known for some of the earliest documented brewing practices and the "*Hymn to Ninkasi*."

Surfactant: A substance used in sanitizers like iodophors to make cell walls permeable, allowing the biocidal agent to penetrate and kill microbes. It also helps spread the sanitizer evenly across surfaces.

Sustainability: Practices aimed at reducing the environmental impact of brewing, such as efficient water management, renewable energy use, and sustainable ingredient sourcing. Sustainability ensures the long-term viability of the brewing industry.

T

Tannins: Tannins are naturally occurring polyphenolic compounds found in barley husks, hops, and certain types of wood. They can contribute to the astringency and bitterness of beer. During the brewing process, excessive extraction of tannins, often due to high temperatures or improper pH levels during mashing and sparging, can result in unwanted harsh flavors and haze formation. Proper management of tannin levels is essential for achieving a balanced flavor profile and clear beer.

Tap: A device used to dispense beer from a keg, typically connected to a CO2 system to maintain pressure and carbonation. Taps come in various designs for different types of kegs and dispensing systems.

Tartness: A sharp, tangy flavor characteristic of sour beers, resulting from the presence of acids like lactic acid.

Tasting Notes: Detailed descriptions of the aroma, flavor, mouthfeel, and appearance of beer, used to track changes over time and identify areas for improvement.

Temperature Accuracy: The precision of maintaining specific temperatures during mashing to ensure proper enzyme activity and optimal starch conversion.

Temperature Control: Maintaining consistent temperatures during mashing, fermentation, and storage to ensure optimal yeast performance and prevent off-flavor.

Temperature-Controlled Barrels: Barrels equipped with systems to precisely control aging temperatures, enhancing consistency and quality.

Terroir: The unique environmental factors, including local microflora, that influence the character of a beer. Terroir is essential to the distinctiveness of Lambic beers from the Senne Valley.

The Scientific Method: A systematic approach to research involving observation, hypothesis formation, experimentation for verfying or falsifying the hypothesis, data analysis with assessment, replicability, and peer review. The scientific method drives innovation and refinement in brewing techniques.

Thermometer: An instrument used to measure the temperature of the mash, boiling wort, and fermentation, essential for monitoring and controlling the brewing process.

Traditional Methods: Brewing techniques and practices passed down through generations, often characterized by their simplicity and reliance on natural processes. Traditional methods are celebrated for their role in preserving the history and culture of brewing.

Trisodium Phosphate (TSP): A robust cleaning agent used for heavy-duty cleaning tasks in brewing. It is effective in removing heavy soils and residues from brewing equipment.

Two-Row Brewers' Malt: Malt made from two-row barley, preferred for its lower protein and tannin content, resulting in clearer, less astringent beers.

U

Under-Pitching: Adding too little yeast to the wort, which can stress the yeast and lead to slow fermentation and undesirable off-flavors.

Utilization: The effectiveness of hops in imparting their desired qualities, such as bitterness, flavor, and aroma, into the beer.

V

Vacuum Distillation: A technique used to remove alcohol from beer by lowering the boiling point of alcohol through a vacuum, preserving the beer's flavor while making it non-alcoholic.

Vanillin: A flavor compound found in oak that imparts vanilla-like aromas and flavors to beer. It adds a sweet, creamy note, particularly appealing in darker, richer beer styles.

Volatility: The tendency of a substance to vaporize. In hops, volatile oils contribute to aroma and are best preserved through dry hopping.

Vorlof (Recirculation): The initial step in lautering where wort is recirculated through the grain bed to clarify it before being drained to the boil kettle.

W

Water Chemistry: The study of the chemical composition of water, including pH levels and mineral content, and its effects on brewing.

Water Management: Efficient use and recycling of water in brewing operations to reduce consumption and environmental impact. Water management is a key component of sustainable brewing practices.

Water Profile: The specific composition of minerals and pH in a water source, which influences the brewing process and the final beer.

Water Recycling: The process of treating and reusing water within the brewery to reduce overall water consumption. Advanced systems like membrane bioreactors help breweries recycle process water for non-potable uses.

Water Testing: The practice of regularly analyzing water for pH and mineral content to make informed adjustments for brewing.

Water-to-Grain Ratio: The ratio of water to crushed malt in the mash, affecting its consistency and efficiency, commonly around 1.25 to 1.5 quarts of water per pound of grain.

Well-Modified Malts: Well-modified malts are barley malts that have undergone thorough malting, breaking down internal grain structures to convert proteins and starches into soluble forms. This makes them more accessible during mashing, improving extract efficiency and enzyme activity. Well-modified malts require less intensive mashing techniques, leading to smoother fermentation and better beer clarity and stability. They are preferred for producing high-quality ales and lagers due to their ease of use and reliability.

Wheat: An adjunct used to improve head retention and mouthfeel in beers, commonly found in wheat beers and Belgian-style witbiers.

Whirlfloc Tablets: Whirlfloc tablets are a type of fining agent used to clarify wort. Made from a blend of Irish moss (carrageenan) and purified seaweed, these tablets help coagulate proteins and other suspended solids, causing them to settle quickly during the whirlpool stage. This results in clearer wort and, ultimately, clearer beer. Whirlfloc tablets are typically added during the last 15 minutes of the boil. They are valued for their efficiency and ease of use, improving the clarity and quality of the final product.

Whirlpooling: A technique used at the end of the boil where the wort is stirred to create a vortex, helping to separate and remove trub (solid particles) from the wort before fermentation.

Wild Yeast: Naturally occurring yeast strains that contribute unique flavors and characteristics to beer. Wild yeast is used in spontaneous and mixed fermentation processes.

Wind Turbines: Devices that convert wind energy into electricity, providing a renewable energy source for breweries in suitable locations. Wind turbines help reduce reliance on fossil fuels and lower greenhouse gas emissions.

Witbiers: A traditional Belgian-style beer, also known as Belgian White, brewed with a significant portion of wheat in addition to barley malt. Witbiers are typically unfiltered, giving them a cloudy appearance. They are often spiced with coriander and orange peel, resulting in a refreshing, mildly spiced flavor profile with citrus notes. These beers are usually light-bodied with a pale, straw-like color and a crisp, slightly tart finish.

Wood Chips: Small pieces of oak or other woods added directly to the fermenter or conditioning vessel to impart wood flavors to beer. They provide quick flavor extraction due to their high surface area.

Wood Spirals: Coiled pieces of wood that provide a larger surface area for flavor extraction. They offer a controlled and gradual release of flavors, mimicking the effects of barrel aging in a more manageable format.

Woodruff: A flavoring syrup traditionally used in Berliner Weisse to balance its tartness. It adds a sweet, herbal note to the beer.

Workshops and Conferences: Events where brewers can learn from industry experts, participate in practical demonstrations, and network with fellow brewers to enhance their skills and knowledge.

Wort: The sweet liquid extracted from the mash during lautering, containing fermentable sugars that yeast will ferment to produce beer. The wort pH should be between 5.0-5.4 to avoid harsh bitterness and achieve the best flavor from the hops during the boil.

Wort Clarity: The absence of grain particles and other solids in the wort, essential for producing clear, high-quality beer.

Wort Concentration: The process of evaporating excess water during boiling to increase the wort's gravity, crucial for controlling the beer's final alcohol content and body. The wort pH should be between 5.0-5.4 to avoid harsh bitterness and achieve the best flavor from the hops during the boil.

Y

Yeast: Microorganisms used in brewing to ferment wort, converting sugars into alcohol and carbon dioxide.

Yeast Health: The vitality and viability of yeast, essential for a vigorous and complete fermentation.

Yeast Management: Practices that ensure healthy and effective fermentation, including proper pitching rates, aeration, temperature control, and sanitation.

Yeast Starter: In brewing, a yeast starter is a small batch of wort that is used to propagate yeast cells before pitching them into the main batch of beer. This process involves mixing water, malt extract, and yeast, allowing the yeast to multiply and become more active. A yeast starter helps ensure a strong and healthy fermentation by increasing the yeast cell count and vitality, leading to better attenuation and fewer off-flavors. Starters are especially useful for high-gravity beers, lagers, or when using older yeast. Proper aeration and nutrient addition are essential to optimize yeast growth in the starter.

Yeast Washing: In brewing, yeast washing is a process used to clean and purify harvested yeast for reuse in future batches. This technique involves separating the yeast from trub (protein, hop residue, and other debris) and unwanted microorganisms. The process typically includes adding sterile water or a dilute acid solution to a yeast slurry, allowing the mixture to settle, and then decanting the clean yeast from the top layer. Yeast washing helps maintain yeast vitality and health, ensuring consistent fermentation performance and preventing off-flavors. Proper sanitation and careful handling are crucial to avoid contamination during the washing process.

Z

Zymurgy: The science and study of fermentation processes in brewing and winemaking, encompassing the biochemical principles behind brewing.

Why did the barley join the alphabet club? Because it heard they were going to Zymurgy and didn't want to miss out on the fun!

***

Author

Dr. Jeffrey W. Bulger is a passionate brewer who seamlessly blends science, philosophy, and craftsmanship in his approach to brewing. Holding a Ph.D. in Philosophy with a specialization in Biomedical Ethics from the University of Tennessee, he brings a unique perspective to the art and science of beer making. His extensive studies in the philosophy of science and ethics provide a deep understanding of the principles that underlie brewing processes.

Before pursuing philosophy, Dr. Bulger earned a Bachelor's degree in Petroleum Engineering and Geology from the University of North Dakota, grounding his knowledge in the natural sciences. A dedicated homebrewer for decades, he became a certified beer judge through the Beer Judge Certification Program (BJCP) in 2015, reflecting his commitment to excellence and deep appreciation for the craft.

Throughout his career, Dr. Bulger has been dedicated to education and mentorship, guiding students and fostering a love of learning. His interdisciplinary expertise allows him to explore brewing through the lenses of rationalism, empiricism, and the scientific method, offering readers a comprehensive understanding of both traditional techniques and modern innovations.

In "**The Art and Science of Brewing**," Dr. Bulger invites you to join him on a journey that celebrates the fusion of tradition and innovation. Whether you're a novice brewer or a seasoned professional, his insights and guidance will inspire you to elevate your brewing practice and appreciate the rich tapestry of history, science, and creativity that defines the world of beer.

When asked about the key to brewing success, Dr. Bulger smiles and says, “It’s all about balance—in flavors, in methods, and in life!”

***

Notes

Brewing notes are the brewer's compass, guiding each batch with wisdom from past brews and inspiration for future creations.

Notes

Notes

Notes

Why don't breweries ever go out of business? Because they always have a "crafty" plan brewing!

***

Made in the USA
Las Vegas, NV
02 October 2026

7273b71f-8740-43f2-85ae-e1840b304290R01